Harmeet Singh

Drone Dynamics

Engineering, Applications, and Innovations in Aerial Robotics

Table of Contents

Chapter 1: Introduction to Drones 6
1.1 Definition and Overview 6
1.2 Evolution of Aerial Robotics 7
1.3 Importance of Drone Technology in Modern Times 8
1.4 Key Components of a Drone 10
1.5 Overview of Regulations and Ethical Considerations 12

Chapter 2: History of Drone Technology 15
2.1 Early Innovations in Aerial Technology 15
2.2 Military Applications in the 20th Century 16
2.3 The Rise of Civilian and Commercial Drones 18
2.4 Evolution of Control Systems and Autonomy 20
2.5 Key Milestones in Drone Development 21

Chapter 3: Types of Drones 23
3.1 Fixed-Wing Drones 23
3.2 Rotary-Wing Drones (Quadcopters, Helicopters) 24
3.3 Hybrid Drones 27
3.4 Nano and Microdrones 29
3.5 Specialized Drones for Unique Applications 30

Chapter 4: The Science of Flight 32
4.1 Principles of Aerodynamics 32
4.2 Lift, Thrust, Drag, and Weight Dynamics 33
4.3 Stability and Maneuverability in UAVs 35
4.4 Role of Propulsion Systems 37
4.5 Advances in Lightweight Materials 37

Chapter 5: Drone Engineering and Design 39
5.1 Structural Components and Materials 39
5.2 Propulsion Systems and Motors 41
5.3 Power Sources: Batteries, Solar, and Hybrid Systems 43
5.4 Sensors and Navigation Systems 45
5.5 Customizing Drones for Specific Applications 46

Chapter 6: Autonomous Systems and Artificial Intelligence 49
6.1 Understanding Drone Autonomy Levels 49
6.2 Role of Machine Learning in Drone Operations 51
6.3 Computer Vision and Object Recognition 53
6.4 Path Planning and Obstacle Avoidance 54

6.5 Challenges in Developing Fully Autonomous Drones 55

Chapter 7: Communication and Control Systems 58

7.1 Remote Control and Radio Frequencies 58

7.2 Ground Control Stations and Interfaces 59

7.3 Real-Time Data Transmission 61

7.4 Satellite and Cellular Connectivity 63

7.5 Security and Privacy in Communication 64

8.1 Precision Farming and Crop Monitoring 66

8.2 Pest Control and Spraying Systems 68

8.3 Soil and Field Analysis 70

8.4 Livestock Management 71

8.5 Benefits and Challenges in Agri-Tech Adoption 73

Chapter 9: Applications in Delivery and Logistics 76

9.1 Drone-Based Package Delivery Systems 76

9.2 Last-Mile Delivery Solutions 78

9.3 Inventory Management in Warehouses 79

9.4 Disaster Relief and Medical Supply Distribution 81

9.5 Regulatory Hurdles and Public Perception 83

10.1 Role in Law Enforcement 85

10.2 Border Patrol and Coastal Monitoring 86

10.3 Public Safety During Events and Emergencies 88

10.4 Anti-Drone Systems and Countermeasures 89

10.5 Ethical Implications of Surveillance 91

Chapter 11: Environmental and Scientific Research 93

11.1 Monitoring Climate Change and Ecosystems 93

11.2 Wildlife Conservation Efforts 95

11.3 Geological and Mineral Surveys 97

11.4 Atmospheric and Weather Data Collection 98

11.5 Role in Marine and Oceanic Research 100

Chapter 12: Drones in Urban Development 102

12.1 Infrastructure Inspection and Maintenance 102

12.2 Role in Smart Cities 104

12.3 Urban Planning and 3D Mapping 106

12.4 Integration with IoT Systems 108

12.5 Challenges in Urban Drone Deployments 109

Chapter 13: Military and Defense Applications 111

- 13.1 Tactical Reconnaissance and Surveillance 111
- 13.2 Combat and Weaponized UAVs 113
- 13.3 Unmanned Logistics Support 115
- 13.4 Counter-Insurgency and Anti-Terrorism Operations 116
- 13.5 Ethical Concerns in Autonomous Warfare 118

Chapter 14: Innovations in Drone Technology 121
- 14.1 Miniaturization and Swarm Robotics 121
- 14.2 Advances in Battery and Power Systems 123
- 14.3 5G Connectivity and Drone Integration 125
- 14.4 Biologically Inspired Drones 127
- 14.5 Emerging Technologies in UAV Design 128

Chapter 15: Regulations and Legal Framework 131
- 15.1 Global Drone Regulations Overview 131
- 15.2 Airspace Management and Integration 132
- 15.3 Data Privacy and Security Concerns 134
- 15.4 Licensing and Operator Training Requirements 136
- 15.5 Balancing Innovation with Public Safety 138

Chapter 16: Challenges in Drone Technology 140
- 16.1 Technical Challenges in Drone Development 140
- 16.2 Weather and Environmental Limitations 142
- 16.3 Issues with Scaling for Commercial Use 144
- 16.4 Public Acceptance and Trust 146
- 16.5 Addressing Ethical and Privacy Concerns 147

Chapter 17: Future of Drone Technology 149
- 17.1 Trends in Autonomous Systems 149
- 17.2 Potential in Space Exploration 150
- 17.3 Role in Global Connectivity Initiatives 152
- 17.4 Emerging Applications in Healthcare and Education ... 153
- 17.5 Vision for the Next Decade 155

Chapter 18: Case Studies and Success Stories 156
- 18.1 Drones in Disaster Management: Real-World Applications 156
- 18.2 Transformative Impact on Industries 157
- 18.3 Startup Ecosystems and Innovations in UAVs 159
- 18.4 Lessons from Failed Projects 161
- 18.5 Inspiration for Future Developments 162

Preface

The world is undergoing a technological revolution, and at the forefront of this change is the rise of unmanned aerial vehicles (UAVs), more commonly known as drones. From humble beginnings as remote-controlled toys to sophisticated machines capable of transforming entire industries, drones are now a cornerstone of technological innovation. Their impact spans across diverse sectors, from agriculture and logistics to healthcare, defense, and environmental research. Drones are no longer just an idea or a novelty; they are an essential tool in reshaping how we live, work, and interact with our environment.

This book, *Drone Dynamics*, seeks to capture the full scope of the rapidly advancing field of drone technology. It serves as both a comprehensive introduction for those new to the subject and a detailed resource for those looking to understand the complex dynamics behind drone operations, design, and their broad array of applications. With the explosive growth of drone technology, it becomes increasingly important to understand both the technical intricacies and the profound impact these innovations will have on various industries and society as a whole.

The chapters in this book delve deeply into the core aspects of drone technology, beginning with the fundamental principles of aerodynamics and moving through detailed explorations of drone design, autonomous systems, communication technologies, and cutting-edge innovations. From understanding the scientific and engineering foundations of drone operations to exploring real-world applications such as disaster management, agriculture, surveillance, and military use, this book provides a holistic view of how drones are shaping the future.

Throughout *Drone Dynamics*, readers will find a wealth of case studies and success stories illustrating the power of drones to solve complex problems. However, the book does not shy away from discussing the challenges and ethical concerns surrounding the widespread use of UAVs. Issues such as regulatory compliance, privacy, security, and public perception are also explored, offering a balanced perspective on the future of drone technology.

As we look forward to the next decade of drone development, it is clear that drones are set to play an even more integral role in our daily lives. The potential for autonomous systems, innovations in battery technology, and new applications in space exploration and healthcare promise to further revolutionize industries and improve lives in ways we can only begin to imagine. This book aims to provide both the technical foundation and the visionary outlook needed to understand the current state of the drone industry and its boundless possibilities.

Drone Dynamics is written for a wide audience, including engineers, entrepreneurs, researchers, policymakers, and anyone with an interest in the growing field of drone technology. Whether you are looking to deepen your understanding of the science behind drones, explore their many uses, or consider their future impact, this book offers the knowledge and insights to help guide you through this exciting era of unmanned aviation.

I hope this book serves as both an educational resource and an inspiration for those passionate about the power of drones and their potential to shape the world in remarkable ways.

Dr. Harmeet Singh

Chapter 1: Introduction to Drones

1.1 Definition and Overview

Drones, also referred to as unmanned aerial vehicles (UAVs), are aerial machines designed to fly without a pilot on board. These aircraft can be remotely controlled or operated autonomously through computer systems and sensors. Drones come in various shapes and sizes, ranging from small quadcopters with four propellers to large fixed-wing systems used in military and industrial applications. Drones have gained immense popularity in recent years, with applications ranging from recreational use to complex commercial and military operations.

A drone typically comprises several parts, including motors, propellers, sensors, a flight controller, and a power source. In addition, many drones are equipped with cameras and GPS technology, which enables them to capture high-resolution images, navigate autonomously, and return to their starting point when necessary. Their capacity to carry different payloads, such as cameras, sensors, and delivery packages, is one of the reasons why they have become versatile tools for both personal and professional use.

The operation of drones is governed by a combination of hardware and software that ensures safe and efficient flying. A drone's design, control system, and sensors work in tandem to allow the drone to perform various tasks, including surveillance, mapping, surveying, and transportation of goods. The integration of artificial intelligence (AI) and machine learning (ML) has further enhanced drones' autonomy, enabling them to adapt to dynamic environments, make decisions, and perform complex missions with minimal human intervention.

The uses of drones have expanded dramatically over the past two decades. Originally developed for military purposes, drones are now utilized in many sectors, including agriculture, logistics, emergency services, environmental monitoring, filmmaking, and construction. The global commercial drone market is expected to grow exponentially, driven by advancements in technology and the demand for cost-effective, efficient solutions for industries across the globe.

1.2 Evolution of Aerial Robotics

The evolution of aerial robotics, which includes drones, can be traced back to the early 20th century. The first recorded instances of drone-like devices were primarily for military purposes, such as training targets for anti-aircraft gunners. In 1916, during World War I, the concept of a radio-controlled aircraft emerged as a tool for military training. This early version of a drone was rudimentary, but it laid the groundwork for future developments.

Throughout the mid-20th century, the evolution of aerial robotics was driven by military interests, primarily for surveillance and reconnaissance. In the 1960s and 1970s, the U.S. military began using drones for reconnaissance missions in Vietnam, where they served as eyes in the sky, collecting intelligence and surveilling enemy movements. These early UAVs were generally large, fixed-wing aircraft that required a substantial amount of human involvement in their operation.

In the 1990s, the drone industry began to experience rapid innovation with the introduction of smaller, more lightweight UAVs. The development of GPS technology and more powerful batteries allowed drones to become more agile, extend their flight time, and increase their payload capacity. These advancements made it possible for drones to be used in civilian applications, such as aerial photography and surveying.

One of the key milestones in the evolution of drones came in the early 2000s with the development of quadcopters. Unlike their fixed-wing counterparts, quadcopters, which have four rotors, were able to take off and land vertically, making them highly maneuverable and suitable for a variety of applications. The launch of consumer-grade drones, such as the DJI Phantom, brought

UAVs into the mainstream, allowing hobbyists and professionals to use them for recreational flying, photography, and videography.

In the past decade, drones have seen significant improvements in their capabilities, driven by advancements in AI, sensors, and computational power. Autonomous flight systems, obstacle detection and avoidance, and real-time data processing have transformed drones into sophisticated tools that can operate without human intervention in many cases. Moreover, drones have become more affordable and accessible, allowing them to penetrate industries such as agriculture, logistics, search and rescue, and environmental monitoring.

The continued advancement of drone technology, particularly in the areas of battery efficiency, artificial intelligence, and communication systems, suggests that drones will only become more integrated into daily life and business operations in the years to come.

1.3 Importance of Drone Technology in Modern Times

Drone technology has rapidly evolved to become one of the most important innovations in modern times. Drones have found applications across multiple sectors, from agriculture and logistics to environmental conservation, disaster management, and filmmaking. The significance of drones in today's world cannot be overstated, as they have transformed traditional methods of performing tasks, providing greater efficiency, safety, and precision.

In agriculture, drones have revolutionized how farmers manage their crops and livestock. Through the use of aerial imagery and sensors, drones can monitor crop health, assess water levels, detect pests, and apply fertilizers and pesticides with pinpoint accuracy. This not only reduces the need for manual labor but also promotes sustainable farming practices by minimizing the use of chemicals and water.

In logistics, drones have the potential to disrupt traditional delivery methods. Companies like Amazon are actively exploring the use of drones for last-mile delivery, reducing the time it takes to deliver packages from distribution centers to consumers. Drones can access remote and hard-to-reach areas where conventional vehicles cannot, providing a more efficient solution for delivering goods in urban, rural, and isolated locations.

Drones are also playing a critical role in disaster response and search-and-rescue operations. In situations where human access is limited or dangerous, drones can be deployed to provide real-time aerial surveillance, locate missing persons, and deliver supplies to hard-to-reach areas. For example, drones were used extensively during the 2015 earthquake in Nepal to assess the damage and deliver aid to affected communities.

In the realm of environmental conservation, drones have proven invaluable in monitoring wildlife, tracking deforestation, and studying climate change. They allow researchers to gather data in remote locations without disturbing ecosystems. Drones are also used to monitor air quality, track ocean currents, and map environmental changes, providing data that helps inform policies and conservation strategies.

Moreover, drones have transformed industries such as filmmaking and media production. Aerial shots once required expensive helicopters or cranes, but drones have made it possible for filmmakers to capture stunning visuals at a fraction of the cost. The ability to control drones with precision has opened up new creative possibilities in cinematography.

The potential of drones extends far beyond the applications mentioned above. As technology continues to advance, drones are expected to play an even more prominent role in fields such as infrastructure inspection, healthcare, and education. Drones can inspect bridges, pipelines, and communication towers, saving time and improving safety by reducing the need for human workers to perform dangerous tasks at heights. In healthcare, drones are being tested to deliver medical supplies, including vaccines and blood samples, to remote or underserved areas.

As drone technology continues to evolve, it will likely lead to new innovations that further transform industries and society. Drones hold the promise of making many processes faster, safer, and more efficient, all while reducing the environmental impact of traditional methods.

1.4 Key Components of a Drone

A drone's functionality and performance are dependent on several key components working together seamlessly. These components include the airframe, propulsion system, flight controller, sensors, and power supply. Understanding the role and function of each component is essential for understanding how drones operate and perform their various tasks.

Airframe

The airframe is the structure of the drone that houses all of its components. It is typically made from lightweight materials such as carbon fiber, plastic, or aluminum, which help reduce the overall weight of the drone while maintaining strength and durability. The design of the airframe can vary depending on the type of drone, whether it's a quadcopter, hexacopter, or fixed-wing UAV. The airframe must be sturdy enough to withstand the forces of flight and any external conditions (e.g., wind or turbulence) that may be encountered during operation.

Propulsion System

The propulsion system is responsible for providing the thrust necessary for the drone to take off, hover, and move through the air. It typically consists of motors, propellers, and sometimes a fan or jet engine in larger drones. The number of motors varies depending on the drone's design. For example, quadcopters have four motors and propellers, while hexacopters have six. The efficiency and power of the motors directly influence the drone's flight time, stability, and payload capacity. The propellers, which spin at high speeds, generate lift and thrust, and their size and pitch are optimized for maximum efficiency based on the drone's size and purpose.

Flight Controller

The flight controller is the brain of the drone. It is responsible for processing inputs from the pilot (or autonomous control system) and translating them into the appropriate commands that control the motors and other components of the drone. The flight controller uses data from various sensors, including accelerometers, gyroscopes, and magnetometers, to ensure stable flight. It adjusts the speed and direction of the motors to keep the drone balanced and on course. Advanced flight controllers can enable drones to perform automated tasks such as GPS navigation, obstacle avoidance, and autonomous flight.

Sensors

Drones are equipped with a variety of sensors that enable them to interact with the environment, gather data, and navigate autonomously. Some of the most common sensors found on drones include:

1. **GPS:** Global Positioning System (GPS) sensors allow drones to navigate autonomously, follow waypoints, and return to their home location. GPS is crucial for maintaining stable flight and ensuring accurate positioning.
2. **IMU (Inertial Measurement Unit):** An IMU includes accelerometers, gyroscopes, and magnetometers to measure the drone's orientation, speed, and rotation. These sensors help stabilize the drone and keep it level during flight.
3. **Barometer:** A barometer measures air pressure, helping the drone determine its altitude above the ground.

4. **Cameras and LIDAR:** Cameras and Light Detection and Ranging (LIDAR) sensors are used for capturing images, videos, and 3D maps. They are essential for tasks such as aerial photography, surveying, and environmental monitoring.

Power Supply

The power supply of a drone typically consists of a battery, most commonly a lithium-polymer (LiPo) battery. The battery stores the energy needed to power the motors, sensors, and flight controller. The capacity and efficiency of the battery are key factors that determine the drone's flight time and range. Advances in battery technology have helped extend the flight time of drones, though battery life remains one of the main limitations for long-duration flights.

1.5 Overview of Regulations and Ethical Considerations

As drones have become more prevalent in both civilian and commercial sectors, they have raised important legal, regulatory, and ethical issues. The rapid expansion of drone technology has prompted governments around the world to establish frameworks to regulate their use and ensure safety, privacy, and accountability.

Regulatory Frameworks

One of the most significant regulatory bodies for drones is the Federal Aviation Administration (FAA) in the United States. The FAA regulates the operation of drones in national airspace to prevent collisions with manned aircraft and ensure public safety. In addition to the FAA, various countries have their own regulatory bodies that impose guidelines on drone usage. These regulations cover aspects such as registration, operational limits, and certification requirements for drone pilots. For instance, drones used for commercial purposes in the U.S. must be registered with the FAA, and drone operators must hold a Remote Pilot Certificate.

The International Civil Aviation Organization (ICAO) also provides global guidelines for drone operations, with a focus on ensuring that drones do not pose a risk to public safety or aviation systems. National regulatory bodies, such as the European Union Aviation Safety Agency (EASA), have worked to harmonize drone regulations across member states to create a more cohesive global standard for drone use.

Privacy and Ethical Concerns

Drones, particularly those equipped with cameras and sensors, raise significant concerns regarding privacy. The ability of drones to capture high-definition images and videos from the sky has led to fears about surveillance and invasion of privacy. In many countries, regulations have been established to limit drone flights over private property, residential areas, and sensitive locations such as government buildings and military installations.

Ethical considerations also come into play in areas such as drone warfare, environmental monitoring, and wildlife protection. In military applications, drones have been used for targeted strikes, raising questions about the ethics of remote warfare and the potential for civilian casualties. Similarly, drones used for environmental monitoring and wildlife conservation must be operated in a way that minimizes disturbance to ecosystems and animals.

Drone manufacturers and operators are increasingly recognizing the importance of ethical practices and privacy considerations. Many companies are developing features such as geofencing, which prevents drones from flying in restricted areas, and promoting responsible drone use to mitigate privacy concerns.

In conclusion, drone technology has rapidly advanced over the past few decades, becoming a vital tool in a wide range of industries and applications. From their military origins to their current use in agriculture, logistics, filmmaking, and disaster response, drones are reshaping the way we interact with the world around us. As drone technology continues to evolve, it will bring both opportunities and challenges, requiring careful consideration of regulations, safety measures, and ethical implications to ensure that drones are used responsibly and effectively.

Chapter 2: History of Drone Technology

2.1 Early Innovations in Aerial Technology

The origins of drone technology can be traced back to the early 20th century, when the concept of flying machines was still in its infancy. The evolution of aerial technology was driven by the desire for improved reconnaissance, defense, and transport, particularly during times of war. The earliest innovations in aerial technology were predominantly focused on manned aircraft, yet they laid the groundwork for the development of unmanned aerial vehicles (UAVs), or drones, which would emerge several decades later.

In the years leading up to World War I, aviation pioneers such as the Wright brothers were pioneering powered flight. These early aircraft were fragile and limited in their capabilities, but they established the fundamental principles of aerodynamics and flight mechanics. As aviation continued to develop, it became clear that aircraft could serve many purposes beyond transportation, including reconnaissance and surveillance. The need for gathering intelligence during wartime was especially apparent, and early concepts of unmanned flight began to take shape.

During the 1910s and 1920s, the idea of using radio-controlled aircraft for military purposes emerged. In 1916, a precursor to modern drones was developed by Charles Kettering, an engineer for the U.S. Army Signal Corps. Kettering's "Kettering Bug" was an early attempt at creating an aerial vehicle that could deliver explosives to enemy targets. Though the Kettering Bug was not a drone in the modern sense, it was a significant step forward in the idea of using aircraft for remote-controlled operations.

The first operational use of radio-controlled aircraft came during World War II, when the U.S. military developed the Radioplane, a drone used for target practice. The Radioplane was developed by actor and inventor, Dr. Reginald Denny, who had been experimenting with remote-controlled aircraft since the late 1930s. The Radioplane was the world's first successful radio-controlled

UAV, and it marked a turning point in aerial technology. These drones were primarily used as targets for training anti-aircraft gunners but demonstrated the potential for unmanned flight.

Despite these early innovations, drone technology was still in its nascent stages. Most of the technology during this period was focused on military applications, with little attention paid to commercial or civilian use. However, these early experiments were crucial in laying the foundation for the development of more sophisticated drones in the years to come.

2.2 Military Applications in the 20th Century

The 20th century saw dramatic advances in both military aviation and drone technology. Following the end of World War II, a variety of military forces began to explore the potential of unmanned aircraft for reconnaissance, surveillance, and target practice. However, it was during the Cold War that drone technology began to evolve into a tool with serious strategic and tactical value for military operations.

By the 1950s and 1960s, military UAVs were being used for reconnaissance missions, where they could fly over enemy territory to gather intelligence without risking human lives. One of the most notable early drones used for this purpose was the Ryan Model 147, which was introduced in the 1960s. The Model 147 was a small, jet-powered UAV capable of flying high-altitude reconnaissance missions and was used extensively during the Vietnam War. These early reconnaissance drones were used to gather images and intelligence over hostile territory, providing valuable information without the need for manned aircraft, which were at greater risk of being shot down.

As technology advanced, drones began to evolve from simple reconnaissance tools into versatile platforms for a variety of military operations. The development of advanced sensors, including infrared cameras, GPS systems, and radar, allowed UAVs to perform more complex tasks, including surveillance of military targets and communications intercepts. Drones could now operate autonomously, flying predetermined routes and providing real-time data to military commanders. This was a key moment in the evolution of drone technology, as it marked the transition from simple remote-controlled aircraft to sophisticated, autonomous systems capable of completing complex military operations.

During the 1980s and 1990s, military drones began to play an increasingly prominent role in operations such as surveillance and targeting. The use of drones for precision strikes gained particular attention in the 1990s, with the U.S. military deploying UAVs for the first time in combat during the Gulf War. The MQ-1 Predator, a drone developed by General Atomics, became one of the most famous UAVs of this era. Equipped with both surveillance cameras and weapon systems, the Predator allowed military forces to gather real-time intelligence and engage targets with high precision. The success of the MQ-1 Predator in the Gulf War marked a turning point in the use of drones for military operations, establishing them as critical tools in modern warfare.

The U.S. military continued to invest heavily in drone technology throughout the 21st century, particularly after the events of September 11, 2001. The War on Terror saw the rapid deployment of drones for counterterrorism operations, particularly in regions like Afghanistan and Pakistan. UAVs such as the MQ-9 Reaper, an advanced version of the MQ-1 Predator, became central to the U.S. military's strategy, carrying out surveillance and targeted strikes against enemy combatants and terrorist leaders. The ability to operate drones remotely, often from thousands of miles away, raised the strategic value of UAVs in modern military operations, especially as drone technology became more sophisticated, capable of operating in contested and hostile environments.

The military applications of drones have continued to evolve into the present day. Modern UAVs are now equipped with cutting-edge sensors, artificial intelligence (AI), and machine learning algorithms that allow them to operate autonomously, identify targets, and perform complex missions with minimal human intervention. Drones are now used not only for reconnaissance and targeted strikes but also for electronic warfare, communications, and logistics support.

2.3 The Rise of Civilian and Commercial Drones

While drones began as primarily military tools, the development of smaller, more affordable UAVs in the late 20th and early 21st centuries led to their adoption in civilian and commercial applications. The commercial drone industry began to take shape in the 2000s, fueled by advancements in technology, miniaturization, and the growing availability of GPS, sensors, and cameras.

The early civilian drones were used primarily for hobbies and recreational purposes. In the mid-2000s, companies such as Parrot and DJI began to release affordable consumer-grade drones that were easy to fly and equipped with cameras for aerial photography and videography. These drones were small, lightweight, and could be operated by anyone with a basic understanding of remote controls. As drones became more accessible to the general public, they quickly gained popularity among hobbyists, filmmakers, and photographers who saw the potential to capture aerial footage without the need for expensive helicopters or airplanes.

However, it wasn't long before the commercial potential of drones began to be recognized. In the early 2010s, businesses in a wide range of industries began to experiment with drones to perform tasks that had traditionally been done by humans or larger, more expensive aircraft. In agriculture, drones were quickly adopted for precision farming, where they could be used to monitor crop health, assess irrigation levels, and apply pesticides or fertilizers. The use of drones in agriculture allowed farmers to reduce costs and increase efficiency, all while improving sustainability and minimizing environmental impact.

The logistics industry also began to explore the use of drones for delivery. Companies such as Amazon and Google's Project Wing began testing drone delivery systems, aiming to provide fast, efficient, and cost-effective solutions for last-mile deliveries. In 2013, Amazon introduced the concept of "Prime Air," a service that would use drones to deliver packages within 30 minutes of an order being placed. Although regulatory hurdles and technical challenges have delayed the widespread use of delivery drones, the concept sparked significant interest and investment in the development of UAVs for logistics and e-commerce.

Drones also found applications in industries such as construction, surveying, environmental monitoring, and emergency services. In construction, drones could be used to survey sites, monitor progress, and inspect buildings for damage. In surveying, drones provided a cost-effective and efficient method for creating high-resolution maps and 3D models of terrain. Environmental organizations began using drones to monitor wildlife populations, track deforestation, and assess the health of ecosystems. Emergency services, including search and rescue teams, also adopted drones for their ability to quickly access hard-to-reach areas and provide real-time imagery to aid in decision-making.

In addition to these industries, drones became a common tool for filmmaking and media production. Aerial footage, once limited to expensive helicopter shots, became more affordable and accessible thanks to consumer drones with high-definition cameras. Filmmakers began using drones to capture stunning visuals and cinematic shots, allowing for more dynamic and creative storytelling.

By the mid-2010s, the drone industry had reached a tipping point, with both consumer and commercial drones becoming an integral part of many industries. This period marked the beginning of rapid growth and expansion in the drone market, and the development of increasingly sophisticated UAVs that were capable of performing complex tasks autonomously.

2.4 Evolution of Control Systems and Autonomy

The development of drone control systems and the evolution of autonomy have been key factors in the success and proliferation of drones. Early drones were remote-controlled, meaning that an operator was required to manually control the drone's flight and operations. However, as technology advanced, drones began to incorporate increasingly sophisticated control systems, enabling them to perform more complex tasks with minimal human intervention.

In the 1990s and early 2000s, the introduction of GPS and flight stabilization systems helped pave the way for greater autonomy in drones. These early systems allowed drones to follow predetermined flight paths and maintain stable flight even in turbulent conditions. However, the drones still required constant input from human operators, and their functionality was limited by the capacity of the control systems.

The development of more advanced flight controllers and the integration of artificial intelligence (AI) and machine learning algorithms revolutionized the drone industry. Modern drones are equipped with autonomous flight systems that can navigate complex environments, avoid obstacles, and perform a variety of tasks without human intervention. For example, many drones are now capable of following GPS coordinates, mapping terrain, and even detecting and avoiding obstacles in real time. The introduction of computer vision and lidar sensors further enhanced the capabilities of drones, allowing them to operate autonomously in environments with minimal human oversight.

Autonomous drones have been especially important in industries like agriculture, where drones are required to operate in large, complex environments. The ability to fly autonomously allows drones to monitor crops, survey land, and perform other tasks with high precision, reducing the need for human pilots. Similarly, in industries such as logistics, drones can navigate delivery routes, drop off packages, and return to their base without requiring constant human control.

The evolution of autonomy in drones has not only improved their performance but also expanded their applications. Drones are now capable of performing complex missions, such as inspecting infrastructure, surveying land, and conducting search-and-rescue operations. Autonomous control systems have made it possible for drones to operate in dangerous or inaccessible locations, reducing the risk to human workers and increasing efficiency.

2.5 Key Milestones in Drone Development

The development of drone technology has been marked by a series of key milestones that have shaped the industry and transformed the way drones are used. Some of these milestones include major technological advancements, breakthroughs in autonomy, and significant shifts in the regulatory landscape.

One of the earliest milestones in drone development was the creation of the Kettering Bug during World War I, which was one of the first recorded uses of an unmanned aircraft for military purposes. This was followed by the introduction of the Radioplane during World War II, which marked the beginning of the use of radio-controlled aircraft for target practice.

In the 1960s, the development of the Ryan Model 147 and the use of drones in the Vietnam War marked the beginning of drones' role in military reconnaissance. The 1990s saw the introduction of the MQ-1 Predator, which demonstrated the military value of drones for surveillance and precision targeting.

The civilian drone industry began to take off in the early 2000s with the release of affordable, consumer-grade drones by companies like DJI and Parrot. This democratization of drone technology opened up new possibilities in filmmaking, photography, and various commercial applications.

Perhaps one of the most significant milestones came in the 2010s with the rise of autonomous drones, capable of performing complex tasks with little to no human intervention. The introduction of AI, machine learning, and advanced sensors allowed drones to operate autonomously in dynamic environments, greatly expanding their potential applications.

The 2010s also saw significant changes in the regulatory landscape, with countries around the world beginning to establish frameworks to govern drone use. The Federal Aviation Administration (FAA) in the U.S., for example, implemented regulations that allowed for the safe integration of drones into the national airspace system, paving the way for commercial drone operations.

These milestones, among others, have helped shape the drone industry and continue to drive its growth. As technology continues to advance, drones are expected to play an increasingly important role in industries ranging from logistics and agriculture to search-and-rescue operations and environmental monitoring. The development of autonomous drones, in particular, promises to revolutionize the way we use UAVs in the coming years.

Chapter 3: Types of Drones

3.1 Fixed-Wing Drones

Fixed-wing drones represent one of the most traditional and reliable types of unmanned aerial vehicles (UAVs). Inspired by the design of conventional aircraft, fixed-wing drones use rigid wings to generate lift, which enables them to remain airborne for longer periods of time than most other drone types. These drones are best suited for long-range applications due to their efficiency, and they are often used in a variety of industries, including agriculture, environmental monitoring, mapping, and military operations.

The fundamental design of a fixed-wing drone consists of a fuselage, wings, and a tail, which help maintain stability during flight. Fixed-wing drones operate similarly to traditional airplanes, where the wings generate lift as the drone moves forward. The drone's engines, typically located in the fuselage, provide thrust, while control surfaces such as ailerons, elevators, and rudders are used to adjust the drone's orientation and direction. The flight path of a fixed-wing drone is controlled by altering the position of these surfaces in relation to the air passing over the wings.

One of the primary advantages of fixed-wing drones is their ability to fly long distances without requiring frequent recharging or refueling. The aerodynamic design of the wings reduces drag and allows the drone to maintain stable flight for extended periods of time, sometimes exceeding several hours. This makes them ideal for applications such as aerial surveying, remote sensing, and environmental monitoring. In agriculture, fixed-wing drones are commonly used for crop surveillance over large areas, as they can cover vast expanses more efficiently than rotary-wing drones.

Another key advantage is their speed. Fixed-wing drones can fly at higher speeds than rotary-wing drones, making them more effective for tasks that require high-speed travel over long distances. They are also more energy-efficient, as they do not require the vertical lift that rotary-wing drones rely on. However, fixed-wing drones have a few limitations. For instance, they need to take off and land on a runway or a large open space, as opposed to rotary-wing drones, which can take off and land vertically. This requirement restricts the operational environments in which fixed-wing drones can be deployed.

Fixed-wing drones are also more vulnerable to sudden changes in weather, such as gusty winds or storms, which can affect their stability and flight performance. The relatively large size of many fixed-wing UAVs can also make them difficult to transport and deploy in confined or difficult-to-access locations.

Despite these challenges, fixed-wing drones remain indispensable in many sectors, including agriculture, surveying, and search and rescue. Their long endurance, efficiency, and ability to cover large areas make them the preferred choice for many industrial applications, where high endurance and long-range capabilities are critical.

3.2 Rotary-Wing Drones (Quadcopters, Helicopters)

Rotary-wing drones are another popular category of UAVs, known for their ability to take off and land vertically, offering flexibility in operation that fixed-wing drones cannot match. These drones use rotors to generate lift, with most designs featuring one or more sets of rotors positioned on the drone's body. There are several types of rotary-wing drones, including quadcopters, helicopters, and hexacopters, each of which offers specific advantages for different applications.

Quadcopters

Quadcopters are one of the most common and widely recognized types of rotary-wing drones. As their name suggests, these drones are equipped with four rotors, typically arranged in a symmetrical pattern. The quadcopter's design is simple, compact, and relatively easy to control, making it popular for consumer, commercial, and industrial use. Quadcopters are typically powered by electric motors, though some variations use hybrid systems that combine gas and electric engines to improve endurance and payload capacity.

The quadcopter's four rotors provide the necessary lift for flight, with each rotor spinning in a specific direction to counteract the forces generated by the other rotors. The two front and two rear rotors spin in opposite directions, which helps to stabilize the drone and prevent it from spinning uncontrollably. The pitch, roll, and yaw of the quadcopter are controlled by adjusting the speed of the individual rotors. This system provides a high degree of control over the drone's movement, making it ideal for applications that require precise maneuverability, such as aerial photography, inspection, and surveillance.

Quadcopters are known for their ease of use, stability, and relatively low cost, making them the go-to choice for hobbyists, filmmakers, and commercial applications. With the integration of high-definition cameras and GPS systems, quadcopters have become a powerful tool for capturing aerial footage, surveying land, and inspecting infrastructure. The popularity of quadcopters has also led to the development of a thriving consumer market, with drones available at various price points to suit different needs and skill levels.

One of the most significant advantages of quadcopters is their ability to hover in place, which is particularly useful for tasks that require stationary flight. This capability has made quadcopters indispensable for aerial videography, where capturing smooth, stable shots is essential. They are also used in search-and-rescue operations, where hovering above a specific area can help search teams locate missing persons or assess disaster sites.

Helicopter Drones

Helicopter drones, or single-rotor UAVs, are another type of rotary-wing drone that uses a single, large rotor to generate lift, along with a smaller tail rotor to provide stability and control. These drones operate similarly to traditional helicopters, with the main rotor providing lift and thrust, while the tail rotor prevents the body from spinning uncontrollably. Helicopter drones are typically more efficient than quadcopters, as they are able to generate more lift with fewer rotors, resulting in better flight endurance and a higher payload capacity.

Helicopter drones are often used in more industrial and professional applications, such as aerial surveying, mapping, and payload delivery. Their ability to carry heavier payloads compared to quadcopters makes them ideal for situations where additional equipment, such as cameras or sensors, needs to be deployed from the air. Additionally, single-rotor UAVs are typically more energy-efficient, allowing them to stay in the air longer and cover greater distances.

Despite their advantages, helicopter drones tend to be more complex and expensive than quadcopters. They also require more expertise to operate, as the controls are more intricate, and maintaining stability during flight requires more precision. However, for specialized applications that require longer flight times or heavier payloads, helicopter drones are often the preferred choice.

3.3 Hybrid Drones

Hybrid drones combine the best features of both fixed-wing and rotary-wing designs, offering the versatility of vertical takeoff and landing (VTOL) with the long-range and endurance capabilities of fixed-wing aircraft. Hybrid drones are often referred to as "VTOL aircraft" because they are capable of transitioning between vertical and horizontal flight modes. This makes them ideal for applications that require both the ability to operate in confined spaces and the capability to cover long distances.

The most common design of hybrid drones is the tilt-rotor configuration, where the rotors can tilt between vertical and horizontal orientations. This allows the drone to take off and land vertically, like a helicopter or quadcopter, but once airborne, it can transition to horizontal flight, like a fixed-wing aircraft. This unique capability allows hybrid drones to combine the maneuverability and flexibility of rotary-wing drones with the speed and endurance of fixed-wing aircraft, making them suitable for a wide range of applications, including surveying, search and rescue, and environmental monitoring.

Hybrid drones are used primarily in industrial, commercial, and military applications. For instance, in industries such as agriculture, hybrid drones can fly large distances to survey fields, then transition to vertical flight to maneuver around obstacles such as trees or buildings. In military applications, hybrid drones can be used for both reconnaissance missions and transport, providing a versatile tool for tactical operations.

One of the main advantages of hybrid drones is their ability to operate in both confined and open spaces. The ability to take off and land vertically means that hybrid drones do not require a large runway or open area, while the fixed-wing mode allows them to cover long distances without sacrificing speed or endurance. However, hybrid drones tend to be more complex and expensive than purely fixed-wing or rotary-wing drones due to their specialized design and the technology required to enable them to transition between flight modes.

3.4 Nano and Microdrones

Nano and microdrones are the smallest class of UAVs, designed to operate in very tight spaces and perform tasks that require a high degree of maneuverability and discretion. These drones are typically used in niche applications where their small size and low visibility are an advantage. Nano and microdrones are highly portable, lightweight, and capable of flying in environments that are too confined or hazardous for larger drones to operate.

Nano drones are typically small enough to fit in the palm of a hand, and they often weigh less than 100 grams. Microdrones, while slightly larger, still remain exceptionally compact and lightweight. These drones typically use electric motors and advanced battery technologies to achieve flight, although their small size limits their endurance and payload capacity. The primary appeal of nano and microdrones is their ability to access tight spaces and perform tasks such as indoor inspections, surveillance, and reconnaissance in areas where larger drones would be too cumbersome or visible.

In industries such as building inspection, nano and microdrones can be used to fly through narrow spaces, such as ventilation shafts, pipes, or beneath bridges, where humans would not be able to fit. Their small size and lightweight nature make them less likely to cause damage or create a disturbance during operations, making them ideal for delicate environments.

Moreover, the use of nano drones has grown in recreational and entertainment sectors, as their affordability and ease of use allow consumers to explore new ways of capturing aerial footage. These drones are also increasingly being utilized in research, particularly in the fields of biology, environmental science, and engineering, where their ability to access confined or difficult-to-reach locations makes them invaluable.

3.5 Specialized Drones for Unique Applications

Beyond the standard categories of fixed-wing, rotary-wing, hybrid, and nano drones, there are also specialized drones designed to serve unique applications in a variety of industries. These drones are built with specific capabilities, sensors, and features that make them ideal for highly specialized tasks.

One example of specialized drones is the underwater drone, or autonomous underwater vehicle (AUV), which is designed to operate underwater and is used for tasks such as marine research, underwater mapping, and offshore oil and gas exploration. These drones are equipped with waterproof housings, pressure-resistant materials, and sensors capable of navigating through underwater environments. They are commonly used to explore shipwrecks, conduct marine surveys, and monitor oceanic ecosystems.

Another specialized drone category includes drones designed for agricultural applications. These drones are equipped with advanced sensors, including multispectral cameras, thermal imaging, and other remote sensing technology, to monitor crop health, assess soil conditions, and optimize irrigation. The agricultural drone market has grown significantly in recent years, driven by the need for precision farming, where drones play a crucial role in improving crop yield and reducing the environmental impact of farming practices.

Specialized drones are also used in industries such as mining, construction, and disaster response. Drones used for surveying and mapping in mining can assist with site planning, monitoring environmental conditions, and assessing risks. In construction, drones equipped with 3D scanning and imaging technology can create accurate models of buildings and infrastructure. Disaster response teams use drones equipped with thermal imaging cameras to locate survivors in emergency situations or assess damage after natural disasters.

As drone technology continues to advance, new specialized applications continue to emerge. With the integration of advanced sensors, AI, and autonomous control systems, specialized drones are expected to become even more versatile, capable of performing increasingly complex tasks across a range of industries.

The diverse types of drones outlined in this chapter each have their own set of advantages and limitations, depending on the application. From long-range fixed-wing drones to nimble nano drones and specialized models for specific tasks, the world of drones offers a wide array of options tailored to meet the needs of various industries. As technology evolves, these categories will continue to expand, providing even greater opportunities for innovation in both civilian and commercial sectors.

Chapter 4: The Science of Flight

4.1 Principles of Aerodynamics

Aerodynamics is the study of the motion of air and its interaction with solid objects, such as aircraft and drones. The principles of aerodynamics are fundamental to understanding how drones achieve flight, stay aloft, and perform various maneuvers. The science of aerodynamics is built upon a set of core principles that govern the movement of air around the drone's body. These principles include the concepts of airflow, pressure, and forces acting on the aircraft during flight.

In the case of drones, aerodynamics is essential for achieving efficient flight and ensuring stable flight paths. The primary forces that affect a drone during flight are lift, weight (or gravity), thrust, and drag. Each of these forces interacts with the others, and their balance determines how the drone performs in the air.

One of the key aerodynamic principles is Bernoulli's Principle, which describes how the velocity of airflow affects pressure. In the context of a drone, the rotors (whether fixed-wing or rotary) generate air movement that creates areas of low and high pressure above and below the blades. The difference in pressure causes lift, which is what allows the drone to rise off the ground.

The wing of a fixed-wing drone is designed to shape the airflow in such a way that the air moves faster over the top surface and slower beneath it. This difference in airflow speeds creates lower pressure above the wing and higher pressure beneath it, resulting in lift. In rotary-wing drones, the blades rotate and create a similar effect, generating lift through the interaction between the moving air and the rotating blades.

The drag force, another important aerodynamic concept, is the resistance encountered by the drone as it moves through the air. Drag can be caused by several factors, such as the shape and surface texture of the drone, the density of the air, and the speed of the drone. Minimizing drag is essential for improving the efficiency of flight, as it reduces the amount of energy required to maintain flight.

The lift-to-drag ratio is a key performance metric used to evaluate the aerodynamic efficiency of a drone. A higher lift-to-drag ratio means the drone can generate more lift for a given amount of drag, resulting in greater endurance and performance. Aerodynamic efficiency is critical for drone design, particularly in applications that require long endurance, such as aerial surveying and surveillance.

4.2 Lift, Thrust, Drag, and Weight Dynamics

The dynamics of flight are governed by four fundamental forces: lift, thrust, drag, and weight. These forces work together to influence the flight performance of a drone, and the balance between them determines the drone's ability to fly, maneuver, and maintain stability.

Lift

Lift is the upward force that counteracts the weight of the drone and allows it to remain in the air. For both fixed-wing and rotary-wing drones, lift is generated by the movement of air over the wings or rotor blades. The amount of lift generated depends on several factors, including the shape and size of the wings or blades, the speed of the drone, and the density of the air. A key concept related to lift is the angle of attack, which is the angle at which the wing or rotor blade meets the oncoming airflow. A higher angle of attack typically increases lift but also increases drag, so finding the optimal angle is important for efficient flight.

In fixed-wing drones, the wings are designed to create lift through aerodynamic principles, particularly Bernoulli's Principle. In rotary-wing drones, the blades rotate rapidly to generate lift by displacing air. In both cases, lift must be greater than or equal to the weight of the drone for it to stay airborne.

Thrust

Thrust is the force that propels the drone forward and counteracts the drag force. Thrust is produced by the propulsion system of the drone, which can be either a motor driving a rotor or a fixed engine in the case of fixed-wing drones. In rotary-wing drones, the motors spin the rotor blades, pushing air downwards and generating lift. The force produced by these rotors is also responsible for moving the drone in the desired direction.

In fixed-wing drones, thrust is provided by an engine that drives a propeller or by jet engines in more advanced systems. The thrust produced by these systems must overcome drag and provide sufficient forward velocity to maintain airflow over the wings and generate lift. The amount of thrust required for flight depends on the drone's size, weight, and aerodynamic efficiency.

Thrust is also responsible for controlling the drone's speed. For rotary-wing drones, varying the speed of the motors changes the amount of thrust generated by each rotor, which can affect both the altitude and the forward speed of the drone. In fixed-wing drones, the throttle controls the engine power and, consequently, the amount of thrust produced.

Drag

Drag is the aerodynamic resistance that opposes the motion of the drone through the air. It is a force that works against the drone's forward motion and must be overcome by the thrust produced by the propulsion system. Drag is a product of the drone's shape, surface texture, and speed. It can be categorized into two types: parasite drag and induced drag. Parasite drag is the resistance created by the friction between the air and the surface of the drone. Induced drag is the resistance caused by the generation of lift, particularly in fixed-wing aircraft.

Minimizing drag is crucial for optimizing flight performance, as reducing drag increases the drone's efficiency, enabling it to fly longer distances or stay airborne for longer periods of time. Streamlined designs, smooth surfaces, and careful attention to aerodynamics all play a role in reducing drag.

Weight

Weight is the force exerted on the drone by gravity and is a critical factor in flight dynamics. The weight of a drone must be counteracted by the lift force in order for the drone to remain airborne. The weight of the drone is determined by its mass and the force of gravity acting upon it, which is a constant value on Earth. For a drone to achieve flight, the thrust and lift forces must be greater than or equal to the drone's weight.

The relationship between weight, lift, thrust, and drag is crucial to the design of the drone. If the drone's weight is too high, it may require more lift and thrust to stay airborne, which can lead to inefficient flight and reduced battery life. On the other hand, reducing weight by using lightweight materials can improve efficiency but may limit the drone's payload capacity.

4.3 Stability and Maneuverability in UAVs

Stability and maneuverability are two key characteristics that define the performance of a UAV during flight. Stability refers to the ability of the drone to maintain a steady flight path without excessive oscillation or deviation. Maneuverability refers to the drone's ability to change its flight path, altitude, and orientation quickly and accurately.

Stability

Stability in UAVs is achieved by balancing the forces acting on the drone, particularly the lift, weight, thrust, and drag. A stable drone is able to resist external disturbances, such as wind gusts, and return to its original flight path without requiring constant control adjustments. Stability is influenced by the design of the drone, including the placement of the wings, the size of the stabilizers, and the center of gravity.

For fixed-wing drones, stability is often achieved through the design of the wings and tailplane, which work together to resist unwanted movements. In rotary-wing drones, stability is achieved through the use of gyroscopic forces and flight control algorithms that adjust the speed of individual rotors to maintain balance.

A well-balanced drone will have a center of gravity that is positioned near the center of lift and thrust. If the center of gravity is too far forward or too far back, the drone may become unstable, leading to difficulty controlling flight. Stability is essential for ensuring smooth and controlled flight, particularly in applications such as aerial photography, surveying, and inspection.

Maneuverability

Maneuverability refers to the ability of a drone to change its position and orientation quickly and with precision. For fixed-wing drones, maneuverability is achieved by adjusting the control surfaces (ailerons, elevators, and rudders) to change the drone's attitude and flight path. In rotary-wing drones, maneuverability is achieved by adjusting the speed of individual rotors, allowing for precise control over altitude, direction, and orientation.

Maneuverability is critical in many drone applications, such as search and rescue, where the drone needs to navigate through obstacles or fly in tight spaces. It is also important for applications like cinematography, where smooth and controlled movements are necessary to capture stable, high-quality footage. The ability to perform complex maneuvers, such as hovering, sharp turns, and rapid ascents or descents, is a key factor in evaluating the performance of a UAV.

4.4 Role of Propulsion Systems

The propulsion system of a drone is responsible for providing the thrust required to keep the drone airborne and maneuverable. Propulsion systems can vary depending on the design and intended use of the drone. The primary components of a drone's propulsion system include the motors (electric or internal combustion engines), rotors or propellers, and the power source (typically batteries or fuel cells).

In rotary-wing drones, the propulsion system consists of electric motors that drive the rotor blades. The power provided by the motors determines the thrust produced by the blades. In fixed-wing drones, the propulsion system typically includes an internal combustion engine or an electric motor that drives a propeller. The propeller creates thrust by spinning rapidly and pushing air backward, propelling the drone forward.

The efficiency and power of the propulsion system directly affect the drone's flight time, speed, and ability to carry payloads. Advances in battery technology, such as lithium-polymer (LiPo) and lithium-ion (Li-ion) batteries, have significantly improved the efficiency and energy density of drone propulsion systems, allowing drones to stay in the air for longer periods and carry heavier payloads.

In more advanced UAVs, hybrid propulsion systems are also being explored. These systems combine electric motors for vertical takeoff and landing with gas or jet engines for sustained flight, offering both efficiency and the ability to cover longer distances.

4.5 Advances in Lightweight Materials

The development of lightweight materials has played a crucial role in improving drone performance. Lighter drones are more efficient, as they require less thrust and lift to stay airborne, leading to longer flight times and greater endurance. Lightweight materials also allow drones to carry heavier payloads without compromising their flight capabilities.

A wide variety of materials are used in modern drone construction, including carbon fiber, plastic composites, aluminum, and lightweight alloys. Carbon fiber is one of the most popular materials due to its high strength-to-weight ratio, which makes it ideal for drone frames and structural components. Carbon fiber is both lightweight and durable, enabling drones to withstand the stresses of flight while minimizing weight.

Plastic composites, such as fiberglass, are also commonly used in drone construction, as they are inexpensive and offer a good balance of strength and weight. In addition, advances in 3D printing technology have made it possible to create customized parts and components that are lightweight and optimized for specific applications.

The use of lightweight materials not only improves flight efficiency but also enhances the portability and ease of transportation of drones. Lighter drones are easier to handle, deploy, and transport, making them more practical for a wide range of applications.

In conclusion, the science of flight encompasses a complex interplay of aerodynamic principles, dynamic forces, stability considerations, propulsion systems, and the materials used in drone construction. Understanding these fundamental concepts is essential for designing drones that can achieve efficient and stable flight while meeting the specific needs of various applications. As drone technology continues to evolve, further advancements in aerodynamics, propulsion, and materials will continue to improve the performance and capabilities of UAVs across multiple industries.

Chapter 5: Drone Engineering and Design

5.1 Structural Components and Materials

The design and construction of drones rely heavily on the selection of materials and structural components. Drones are often exposed to harsh environmental conditions and high operational stresses, requiring engineering solutions that balance strength, weight, durability, and cost. The structural integrity of a drone ensures not only its flight capabilities but also its safety and longevity during various operations.

At the heart of drone engineering are the frame and body, which provide the foundation for all other components. These structural components support critical elements like propulsion systems, sensors, and navigation units. The choice of materials and the design of these frames play a pivotal role in the overall performance of a drone, especially in terms of weight reduction and aerodynamic efficiency.

Materials Used in Drone Frames

To build an efficient drone, engineers carefully select materials based on specific properties such as weight, strength, durability, and resistance to environmental factors like moisture, temperature, and UV radiation. Common materials used in drone frame construction include:

1. **Carbon Fiber**: This material is widely used in high-performance drones due to its high strength-to-weight ratio. Carbon fiber is significantly lighter than metals such as aluminum, yet it can bear considerable stress without deforming. Additionally, it is resistant to fatigue and corrosion, making it ideal for drones that experience frequent vibrations and outdoor exposure. Carbon fiber's lightness also leads to greater flight efficiency, contributing to longer flight times and higher payload capacities.

2. **Aluminum**: Aluminum is another popular material used in drone frames. It is heavier than carbon fiber but still offers a favorable strength-to-weight ratio. Aluminum is relatively inexpensive, and it can be easily machined, making it suitable for various structural components like arms, landing gear, and motor mounts. Though not as strong as carbon fiber, aluminum provides adequate strength for drones with moderate weight and less demanding applications.

3. **Plastic Composites (e.g., ABS, Nylon)**: Many commercial and hobbyist drones use plastic composites for their frames, particularly when cost is a concern. These materials are relatively lightweight and inexpensive to produce. While they may not provide the same durability or rigidity as carbon fiber or aluminum, plastic composites are effective for lighter drones that do not face extreme operational conditions. Plastics also allow for complex designs that can be easily manufactured using 3D printing techniques, offering greater flexibility in customizing drone parts.

4. **Fiberglass**: Fiberglass is commonly used in drone frames that require a good balance between weight, strength, and flexibility. It is a composite material made from woven glass fibers embedded in a resin matrix. Fiberglass provides durability without significantly increasing the weight of the drone. It is often used in drone components like propellers and battery enclosures, where both strength and lightness are important.

5. **Titanium**: Titanium is used in specialized drones where high strength, low weight, and corrosion resistance are essential. Titanium is highly resistant to corrosion and offers excellent performance under extreme conditions. However, its high cost limits its widespread use to specialized or military-grade drones.

Design Considerations

The design of the drone's frame includes several key factors, such as aerodynamics, center of gravity, modularity, and impact resistance.

Aerodynamics refers to the shape of the drone, which affects how air flows around the drone during flight. A well-designed frame minimizes drag and maximizes lift, allowing the drone to fly more efficiently.

The center of gravity (CG) is another critical design consideration. The ideal placement of the CG is near the center of the drone to maintain balance and stability. If the CG is too far forward or backward, the drone may experience difficulties with maneuverability or stability. Engineers use lightweight materials and modular designs to ensure that components like the flight controller, sensors, and batteries are properly distributed across the drone's frame.

In addition to aerodynamics and balance, drones must also be designed with impact resistance in mind. Whether a drone is intended for industrial use or recreational flying, it must be able to endure occasional crashes or rough landings. Protective features such as foam padding, landing gear, and impact-resistant outer shells are integrated into the design to minimize damage during such events.

5.2 Propulsion Systems and Motors

The propulsion system of a drone is responsible for providing the thrust needed to lift and maneuver the drone. A propulsion system consists of various components, including motors, propellers, electronic speed controllers (ESCs), and related hardware. The choice of propulsion system depends on the size, weight, and intended use of the drone.

Motors

Drone motors are the key drivers of the propulsion system. There are two primary types of motors used in drones: **brushless DC motors (BLDC)** and **brushed DC motors**.

1. **Brushless DC Motors (BLDC)**: Brushless motors are the most commonly used in drones due to their higher efficiency, longer lifespan, and greater reliability. Unlike brushed motors, which rely on physical brushes to transfer electrical current to the motor's rotating part (the rotor), brushless motors use electronic controllers to regulate the flow of electricity. This design eliminates friction, leading to smoother operation, greater torque, and reduced wear. BLDC motors are ideal for high-performance drones, as they can handle greater loads and provide more precise control.

2. **Brushed DC Motors**: Brushed motors are less expensive and simpler to manufacture but suffer from increased friction and wear. They are typically used in smaller, lower-cost drones, where performance demands are less stringent. While brushed motors are less efficient and have a shorter lifespan than brushless motors, they can still be suitable for hobby-grade drones with less demanding flight needs.

Electronic Speed Controllers (ESCs)

The electronic speed controller (ESC) is a crucial component in a drone's propulsion system. It regulates the speed of the motors by controlling the power sent to the motor. ESCs are responsible for adjusting the rotational speed of each motor, allowing for precise control over the drone's movement. In multi-rotor drones (e.g., quadcopters, hexacopters), each motor is connected to its own ESC, which enables independent control of each rotor's speed.

ESCs play a critical role in the drone's stability, as they help to fine-tune the flight characteristics. For example, if a drone experiences a gust of wind that causes it to tilt, the ESCs adjust the speed of the individual motors to bring the drone back to a stable position. This process is part of the flight control system's feedback loop.

Propellers

The propellers of a drone are responsible for generating lift and thrust. They work by creating a difference in air pressure above and below the blades, which results in the generation of lift. Propellers come in various sizes and materials, and the choice of propeller depends on factors such as the drone's size, weight, and intended application.

Larger propellers generate more thrust but require more power to rotate, while smaller propellers are more efficient but generate less lift. Additionally, the number of blades can impact the efficiency of the propellers. Multi-blade propellers (e.g., three or four blades) provide greater thrust but can also increase drag.

5.3 Power Sources: Batteries, Solar, and Hybrid Systems

A drone's power source is crucial to its operation. The most common power sources for drones are batteries, but other alternatives, such as solar and hybrid systems, are also being explored to improve endurance and efficiency.

Batteries

Currently, the most widely used power source for drones is rechargeable lithium-ion (Li-ion) or lithium-polymer (LiPo) batteries. These batteries offer a favorable combination of energy density, weight, and rechargeability, which makes them suitable for powering drones. Li-ion and LiPo batteries are capable of delivering high currents required by the motors and ESCs without adding significant weight.

1. **Lithium-Polymer (LiPo)**: LiPo batteries are known for their high discharge rates, which makes them ideal for applications that demand high power output, such as racing drones or heavy-lifting industrial drones. LiPo batteries are also lighter than Li-ion batteries, which is a key advantage for flight time and efficiency. However, they tend to be

more sensitive to charging conditions and require careful monitoring to avoid overcharging or undercharging, which can lead to damage or safety hazards.

2. **Lithium-Ion (Li-ion)**: Li-ion batteries have a lower discharge rate than LiPo batteries but offer higher energy capacity and longer cycle life. They are often used in drones that require longer flight times or greater endurance. Li-ion batteries are typically more robust and less prone to the dangers of overcharging, making them more suitable for applications that prioritize stability and longevity over high-performance output.

Solar Power

For drones with long endurance requirements, solar power is an emerging solution. Solar-powered drones are equipped with photovoltaic cells that capture sunlight and convert it into electrical energy. This energy is used to recharge the drone's battery during flight, extending its operating time without requiring landing. Solar drones are particularly useful in remote sensing, environmental monitoring, and military surveillance, where continuous operation is often critical.

However, solar power presents challenges, such as the relatively low efficiency of photovoltaic cells and the limited amount of power they can generate compared to traditional batteries. Solar-powered drones are generally suited for specific applications where long endurance and energy independence are prioritized over high-speed or heavy-lifting capabilities.

Hybrid Systems

Hybrid propulsion systems combine traditional fuel-based engines with electric motors and batteries. This hybrid setup allows drones to achieve greater endurance and flexibility by combining the high energy density of fuel with the low emissions and quiet operation of electric motors. Hybrid systems are gaining popularity in large drones, such as delivery drones or UAVs used for industrial inspections, where the need for extended flight times is essential.

Hybrid drones can switch between electric and fuel power based on the flight phase, providing the benefits of both energy types. These systems are still relatively rare but are being actively developed for more specialized applications, where both long flight times and the ability to carry heavier payloads are essential.

5.4 Sensors and Navigation Systems

Sensors and navigation systems are critical for ensuring a drone's autonomous operation and safety during flight. These systems allow drones to collect data, navigate through environments, and avoid obstacles while maintaining stable flight.

Inertial Measurement Units (IMUs)

An inertial measurement unit (IMU) is a sensor that provides information about a drone's orientation and acceleration. It typically consists of accelerometers, gyroscopes, and magnetometers, which detect changes in speed, angular velocity, and orientation. IMUs are essential for stabilizing a drone during flight and ensuring that it maintains its intended trajectory.

By combining data from multiple IMUs, flight controllers can correct for any deviation in the drone's attitude and adjust the speed of the motors to keep the drone stable. IMUs also play a significant role in autonomous navigation, allowing the drone to track its position and adjust its course.

GPS and GNSS Systems

Global Positioning System (GPS) and Global Navigation Satellite System (GNSS) technologies are commonly used in drones for precise navigation. GPS provides real-time positioning data, allowing drones to determine their location with high accuracy. GNSS systems offer similar capabilities, relying on multiple satellite constellations to provide global coverage and better precision in various environments.

GPS is crucial for enabling autonomous flight, including waypoint navigation, return-to-home (RTH) functionality, and geofencing. These systems help drones navigate without the need for direct control, and they are often combined with other sensors, such as IMUs, to ensure stable and accurate positioning.

Vision and Obstacle Avoidance Sensors

Many drones are equipped with additional sensors for obstacle detection and avoidance. These sensors include ultrasonic sensors, infrared sensors, and cameras that help the drone perceive its environment and detect obstacles in real-time. Visual sensors, such as cameras or LIDAR (Light Detection and Ranging), allow the drone to create 3D maps of its surroundings and avoid collisions by adjusting its flight path.

Obstacle avoidance systems are especially important for drones used in urban environments or areas with dense vegetation. By integrating vision systems with IMUs and GPS, drones can navigate autonomously, avoiding obstacles without human intervention.

5.5 Customizing Drones for Specific Applications

The versatility of drones allows them to be customized for a wide range of applications, from aerial photography and environmental monitoring to military reconnaissance and delivery services. Customizing drones for specific applications requires modifying components such as sensors, payload capacity, flight controllers, and propulsion systems to meet the needs of the task at hand.

Aerial Photography and Videography

For aerial photography and videography, drones are often equipped with high-quality cameras, gimbals for stabilization, and long-duration battery systems. These drones are designed to carry heavy payloads while maintaining smooth, stable flight. Gimbal systems, which are used to stabilize the camera, allow the drone to capture high-quality footage, even in turbulent conditions. Additionally, these drones are equipped with GPS and vision sensors to ensure precise flight control and smooth camera operations.

Delivery Drones

Drones used for delivery purposes require a high payload capacity, long flight times, and precise navigation systems. These drones are often equipped with specialized cargo bays and secure payload release mechanisms. The design also takes into account the need for fast and efficient delivery, so their propulsion systems are typically tuned for optimal speed and energy efficiency.

Industrial Drones

Industrial drones are used in sectors such as construction, agriculture, and inspection. These drones may be customized with payloads like thermal cameras, LIDAR sensors, and GPS receivers for surveying, crop monitoring, or infrastructure inspections. They also tend to have reinforced frames to withstand harsher conditions, as well as enhanced stability and payload handling capabilities.

In conclusion, drone engineering and design are multi-faceted disciplines that require careful consideration of structural materials, propulsion systems, power sources, and sensors. Whether for military, commercial, or recreational purposes, each aspect of a drone's design is tailored to meet the specific demands of the application. Advances in materials science, propulsion technology, and sensor integration continue to push the boundaries of what drones can achieve, making them invaluable tools across a wide variety of industries.

Chapter 6: Autonomous Systems and Artificial Intelligence

6.1 Understanding Drone Autonomy Levels

The concept of autonomy in drone systems refers to a drone's ability to perform tasks without direct human intervention. The increasing adoption of autonomous systems in drones has revolutionized industries by enabling applications such as surveying, mapping, delivery, and inspection with minimal human oversight. Autonomy, in the context of drones, is typically described using levels, which range from manual control to full automation. The levels of autonomy for drones, akin to those used in autonomous vehicles, are categorized to help define the operational capabilities and limitations of various drone systems. Understanding these levels is crucial for evaluating the potential of drones in different applications.

Drone autonomy is typically defined in terms of the degree of human involvement in controlling the vehicle. These levels are defined by the International Organization for Standardization (ISO) in a system that categorizes drones from Level 0 (fully manual) to Level 5 (fully autonomous). The levels are typically broken down as follows:

1. **Level 0 (No Automation)**: At this level, the drone is entirely manually controlled by a human operator. The operator is responsible for all decisions, including navigation, speed control, and handling of unforeseen situations. This level is typical for hobbyist drones or simple unmanned aerial vehicles (UAVs).
2. **Level 1 (Driver Assistance)**: At Level 1, the drone may assist the operator by providing basic stabilization or altitude hold functions, but the operator still has to manage navigation and flight path decisions. These drones often use sensors like GPS or altitude sensors to provide assistance in basic control tasks, though full decision-making remains with the operator.

3. **Level 2 (Partial Automation)**: In this stage, drones have the ability to perform some specific tasks autonomously, but the operator is still responsible for overall control and intervention when necessary. For example, the drone may be capable of following a pre-set path or maintaining a certain speed, but it still requires manual input for changes in direction or altitude. Many modern drones used in commercial applications, such as aerial photography or surveying, function at this level.

4. **Level 3 (Conditional Automation)**: At this level, drones can perform certain tasks autonomously, but they still rely on a human operator to take control in specific situations. The drone can navigate itself in most scenarios, including altitude control, obstacle avoidance, and following predefined routes. However, it may require human intervention in complex or unpredictable environments, such as crowded urban areas or unpredictable weather conditions. Drones used for autonomous deliveries or inspections are often designed to operate at this level.

5. **Level 4 (High Automation)**: Drones at Level 4 are capable of full autonomy within specific environments. This means they can fly and operate without human intervention in certain scenarios, such as delivering packages in designated areas or conducting surveys in remote locations. However, they may still need a human operator to take over in case of system failures, emergencies, or extreme conditions. The drone is able to handle most tasks autonomously, such as navigation, collision avoidance, and real-time decision-making, within a defined operational domain.

6. **Level 5 (Full Automation)**: At Level 5, the drone is capable of operating entirely autonomously in any environment and under any condition without the need for human intervention. These drones can make complex decisions, adapt to changing conditions, and continue performing their tasks without input from the operator. Full autonomy includes everything from navigation to decision-making, including

the ability to autonomously change flight paths, avoid dynamic obstacles, and make real-time adjustments to unforeseen circumstances. This level of autonomy is the ultimate goal for the development of fully autonomous drones, but it is still a work in progress for many drone manufacturers and regulators.

The journey toward Level 5 autonomy involves overcoming numerous technical, regulatory, and ethical challenges, as drones must be able to reliably operate in unpredictable and dynamic environments. The ongoing development of artificial intelligence (AI), machine learning, and other autonomous systems is crucial to achieving higher levels of drone autonomy.

6.2 Role of Machine Learning in Drone Operations

Machine learning (ML) plays a pivotal role in enhancing the autonomy of drones, making it possible for them to perform increasingly sophisticated tasks. At its core, machine learning enables drones to learn from data and adapt to new situations without explicit programming for each possible scenario. By leveraging algorithms that analyze data patterns, drones can improve their decision-making capabilities and execute complex tasks with minimal human oversight.

Machine learning applications in drones span a wide variety of functions, such as navigation, object detection, flight control, and data analysis. These algorithms allow drones to autonomously adapt their flight paths, avoid obstacles, and optimize their performance in real-time. The development of machine learning algorithms specifically tailored to drone operations is one of the key factors enabling drones to move beyond basic automation toward higher levels of intelligence and adaptability.

Supervised Learning

One common form of machine learning used in drone operations is supervised learning, where a model is trained using labeled datasets. For example, a drone could be trained to recognize different types of obstacles or objects by feeding it images of those objects, labeled with information about their location and classification. The drone's algorithms then learn to identify similar objects in real-time, enabling it to navigate and avoid obstacles during flight.

Supervised learning is particularly useful for tasks like image recognition, object detection, and geolocation mapping, where large datasets can be used to train models that improve accuracy over time. In drone operations, supervised learning enables the development of precise control algorithms that help drones maintain stability, avoid collisions, and follow predefined paths.

Reinforcement Learning

Reinforcement learning (RL) is another form of machine learning that is being integrated into drone systems to enable autonomous decision-making. RL involves training models through trial and error, where an agent (in this case, the drone) takes actions in an environment and receives feedback based on those actions. The drone uses this feedback to learn optimal strategies for tasks like navigation and task completion.

For example, reinforcement learning can be used to train drones to navigate complex environments, such as urban landscapes or disaster zones, where obstacles and conditions are constantly changing. The drone can learn to make real-time decisions about its path, flight speed, and maneuvering based on the rewards or penalties it receives from the environment.

By constantly interacting with its surroundings and learning from the outcomes of its actions, a drone can improve its ability to navigate and perform tasks autonomously. RL is particularly useful in dynamic and unpredictable environments where pre-programmed solutions are insufficient.

6.3 Computer Vision and Object Recognition

Computer vision is a critical component of autonomous drone systems. It allows drones to process and interpret visual information from their environment, enabling them to "see" and understand their surroundings. With the help of cameras, LIDAR, and other sensors, drones can capture images, videos, and 3D data, which are then analyzed by computer vision algorithms.

Computer vision enables drones to perform tasks such as object recognition, scene reconstruction, and environmental mapping. These capabilities are essential for autonomous operations, such as obstacle avoidance, precision landing, and real-time navigation.

Object Recognition

Object recognition is one of the key capabilities of computer vision that allows drones to identify and classify objects in their environment. For instance, a drone might be equipped with cameras that capture images of the ground or surrounding environment. The images are then processed by machine learning models trained to recognize specific objects, such as people, vehicles, trees, or buildings.

Object recognition is critical in applications like surveillance, where drones need to distinguish between various objects to monitor activity or identify specific targets. Similarly, in precision agriculture, drones can use object recognition to detect plants, crops, or pests, enabling farmers to monitor crop health and growth.

In industrial settings, object recognition allows drones to detect anomalies in structures or infrastructure. For example, drones equipped with thermal imaging cameras can identify heat signatures of faulty equipment or detect potential hazards, such as overheating machinery or gas leaks.

SLAM (Simultaneous Localization and Mapping)

SLAM is a computer vision technique used by drones to build real-time maps of their environment while simultaneously determining their position within that map. Using a combination of camera data, sensors, and computational algorithms, SLAM allows drones to navigate and operate autonomously in environments where GPS may not be available, such as indoor spaces or urban canyons.

By continuously updating its internal map based on sensor data, SLAM enables drones to perform complex navigation tasks without human intervention. This is particularly useful for drones used in areas with dynamic obstacles or changing conditions.

6.4 Path Planning and Obstacle Avoidance

Path planning is the process by which a drone determines the optimal route to its destination while avoiding obstacles along the way. This process is a crucial aspect of autonomous flight, as it ensures that drones can safely navigate through complex environments. Path planning algorithms take into account various factors such as distance, obstacles, no-fly zones, and dynamic changes in the environment.

Algorithms for Path Planning

Various algorithms are used for path planning in drones. One common approach is the *A algorithm**, which is used to find the shortest path between two points while considering obstacles. The algorithm works by evaluating different potential paths and selecting the most efficient one. It is particularly effective in static environments where obstacles do not change dynamically.

For dynamic environments, drones often use **Rapidly-exploring Random Trees (RRT)** or **Dynamic Programming** techniques to calculate paths in real-time. These algorithms allow drones to continuously update their flight path as new obstacles or changes in the environment are detected. This is particularly important in unpredictable or crowded environments, such as urban areas or disaster zones.

Obstacle Avoidance

Obstacle avoidance is a key challenge in drone autonomy, especially in environments with unpredictable obstacles. Drones are equipped with various sensors, such as ultrasonic sensors, infrared sensors, and cameras, to detect and avoid obstacles. These sensors gather real-time data about the drone's surroundings, which is then processed by the drone's onboard computer to make decisions about navigation.

For example, a drone flying through a forest might use ultrasonic sensors to detect trees and other obstacles in its path. If an obstacle is detected, the drone's control system will adjust its flight path to avoid a collision. Advanced drones may combine obstacle avoidance algorithms with computer vision systems to create a more detailed understanding of the environment, allowing them to identify and avoid objects more accurately.

6.5 Challenges in Developing Fully Autonomous Drones

Despite the significant progress made in drone autonomy, several challenges remain in the development of fully autonomous drones. These challenges span technical, regulatory, ethical, and environmental domains.

Technical Challenges

One of the primary technical challenges is improving the accuracy and reliability of the sensors and algorithms used for autonomous navigation. While current technologies like GPS, cameras, and LIDAR provide valuable data for navigation, they are still prone to errors or failures, particularly in challenging environments such as dense urban areas or indoor spaces. Ensuring that drones can operate reliably in all conditions without human oversight requires overcoming significant technical hurdles related to sensor fusion, real-time data processing, and robust decision-making algorithms.

Regulatory and Safety Concerns

Regulation is another major challenge facing the widespread adoption of fully autonomous drones. Governments around the world are still working to develop clear guidelines for autonomous drone operations, particularly regarding safety and airspace management. The ability for drones to safely share airspace with manned aircraft and other UAVs is a major concern for regulators.

Additionally, there are concerns about the potential misuse of autonomous drones, such as their use for surveillance or other malicious activities. This requires the development of stringent regulations to ensure the ethical use of drones while preventing unauthorized access or control.

Ethical and Social Implications

The ethical challenges of fully autonomous drones are also significant. For instance, the use of drones in surveillance, military operations, or law enforcement raises concerns about privacy, accountability, and transparency. As drones become more autonomous, questions about liability and decision-making also arise, particularly in situations where drones are required to make life-and-death decisions, such as in military operations or search-and-rescue missions.

Environmental Factors

Operating drones in complex and dynamic environments poses significant challenges. Factors such as weather conditions, GPS signal interference, and unforeseen obstacles can disrupt a drone's autonomous flight. Drones must be able to adapt to these changing conditions in real time, making them capable of reliably performing tasks even in environments where human pilots would struggle.

In conclusion, the development of fully autonomous drones requires advances in machine learning, computer vision, sensor technologies, and path planning algorithms. While significant progress has been made, challenges related to technical reliability, regulation, ethics, and environmental conditions must be addressed before fully autonomous drones become a common part of everyday life.

Chapter 7: Communication and Control Systems

7.1 Remote Control and Radio Frequencies

The communication system of a drone is one of its most critical components, as it allows for both control and data transmission between the drone and the operator. The primary mechanism for controlling a drone remotely is through radio frequency (RF) communication. This system facilitates the transmission of commands from the ground station or controller to the drone, allowing the operator to direct its movements, adjust altitude, and activate other functions.

Remote control via radio frequencies operates by using a transmitter (in the operator's hand) and a receiver (on the drone). The operator's controller sends a specific signal, which the receiver on the drone picks up, interpreting it as commands. These commands instruct the drone's flight controller to execute actions such as turning, ascending, or descending.

The use of radio frequencies (RF) for remote control communication offers several advantages, including relatively low latency, high reliability, and the ability to cover significant distances. The selection of appropriate frequencies is critical for ensuring smooth operation, especially in terms of range, interference, and signal strength.

Different types of radio frequencies can be used for drone communication, typically falling within the following bands:

1. **2.4 GHz Frequency Band**: This is one of the most commonly used frequency ranges for drone communication. It offers a good balance of range and data transfer rate while minimizing interference from other devices. The 2.4 GHz band is shared by various consumer electronics, such as Wi-Fi routers, Bluetooth devices, and microwave ovens, which can lead to signal congestion in environments with high electromagnetic activity.

2. **5.8 GHz Frequency Band**: This band is often used for high-definition video transmission and is less crowded than the 2.4 GHz band. However, its range is somewhat limited compared to the lower frequencies, as higher frequencies are more susceptible to attenuation from obstacles, weather conditions, and interference.

3. **Other RF Bands**: For long-range communication or specialized applications, drones can operate in other frequency bands, such as 433 MHz or even proprietary bands used by specific drone manufacturers. These frequencies may provide better signal penetration and longer range but often require more sophisticated equipment and are subject to regulatory constraints in various regions.

Effective communication is essential for drone operations, especially in remote or complex environments, where clear and uninterrupted signals are critical for the safe and reliable operation of the UAV. The signal range, interference resistance, and environmental factors (such as weather or physical obstructions) all influence the choice of frequency.

7.2 Ground Control Stations and Interfaces

Ground control stations (GCS) are the hubs that allow drone operators to interact with and monitor their drones during flight. The GCS is the interface where the operator inputs commands and receives real-time data, such as the drone's position, altitude, battery status, and video feed. The complexity and functionality of ground control stations vary depending on the type of drone and its intended application. Commercial drones for inspection, surveying, and mapping often employ sophisticated ground stations with advanced features, while consumer drones typically use simpler controllers or smartphone apps.

The core function of a ground control station is to transmit commands to the drone and to receive telemetry data back from it. The communication between the GCS and the drone is typically achieved via radio signals, using RF or other communication technologies. More advanced GCS setups might also include satellite communication or cellular connectivity, which can allow drones to operate over much longer distances or in areas where direct RF communication is not feasible.

In terms of the user interface, modern ground control stations often feature graphical displays that show the drone's location, flight path, and any potential hazards, often using GPS mapping systems or live video feeds from the drone's cameras. These displays provide vital information for operators, helping them make informed decisions and ensuring safe operation. Some systems also offer advanced control options, such as autopilot functionality, flight planning, and autonomous mission execution, allowing the drone to follow predetermined routes or complete specific tasks with minimal operator input.

For advanced industrial or military drones, the ground control stations may be much more complex, featuring multiple screens and control systems for handling a variety of sensors, payloads, and communication channels. For example, a drone used for search and rescue missions may be equipped with thermal imaging cameras, and the GCS would need to integrate the live video feed from these cameras along with real-time GPS data, environmental data, and communication links to ground teams.

In addition to hardware controls, the software interfaces of GCS play a crucial role. These interfaces allow for flight planning, execution, monitoring, and real-time troubleshooting. Flight planning software can help operators define waypoints, altitudes, and mission objectives, while real-time telemetry data allows operators to monitor the status of the drone throughout its mission.

A key aspect of ground control stations is the ease with which the operator can interact with the drone and how quickly they can respond to changing conditions. The GCS must be intuitive enough to allow operators to focus on mission execution, rather than struggling with complex control systems. As drones become more autonomous, the role of the GCS may evolve to include mission monitoring and oversight, while drones handle more of the decision-making on their own.

7.3 Real-Time Data Transmission

One of the most important aspects of drone communication systems is the real-time transmission of data between the drone and its ground control station. In many drone applications, particularly in areas like surveillance, aerial photography, infrastructure inspection, and search and rescue, the ability to transmit data in real-time is critical for operational success.

Drones are often equipped with high-definition cameras, sensors, and telemetry systems that collect large amounts of data during flight. This data must be transmitted back to the operator for analysis, decision-making, and mission adjustments. Real-time data transmission enables operators to monitor the drone's status, evaluate environmental conditions, and make immediate adjustments to the flight plan if necessary.

The real-time data transmission process typically involves two main types of communication: **telemetry** and **video feed**.

1. **Telemetry**: This refers to the transmission of flight data, such as position, altitude, speed, battery level, GPS coordinates, and other sensor data. Telemetry data helps the operator track the status of the drone in real-time, ensuring that it is functioning properly and within operational limits. Telemetry systems usually operate on lower bandwidth frequencies, as the data involved is relatively small and periodic.
2. **Video Feed**: Many drones are equipped with cameras for surveillance, inspection, or mapping. The video feed can be transmitted live to the operator, providing visual feedback that is essential for tasks such as visual line-of-sight flying or capturing aerial imagery. Advanced drones often send high-definition (HD) or even 4K video, which requires high-bandwidth communication channels to transmit.

There are several technologies and communication protocols used for real-time data transmission in drones:

1. **Wi-Fi**: Many consumer drones use Wi-Fi for real-time video streaming and telemetry data transmission. While Wi-Fi is convenient and commonly available, its range is limited, making it suitable for short-range operations.

2. **RF Communication**: Drones that require longer ranges, such as those used for surveying, inspection, or delivery, often use radio frequency (RF) communication for data transmission. RF offers greater range and stability, especially in the 2.4 GHz and 5.8 GHz bands.

3. **4G/LTE/5G**: As drone technology progresses, mobile networks such as 4G LTE and 5G are becoming increasingly important for real-time data transmission, particularly in applications like remote inspection, mapping, and delivery. These cellular networks offer a much broader range and bandwidth than traditional RF communication, allowing drones to transmit large volumes of data over long distances.

4. **Satellite Communication (Satcom)**: In remote or inaccessible areas, where traditional communication networks may not be available, drones can rely on satellite communication to transmit data back to the ground control station. Satellite links allow for global connectivity and enable drones to operate in regions with little or no cellular infrastructure.

The development of real-time data transmission systems is driven by the need for higher bandwidth, lower latency, and greater reliability. With increasing demand for high-resolution video feeds, live telemetry, and large-scale data collection, drone manufacturers are working to enhance communication technologies, ensuring that drones can transmit large amounts of data in real-time without interference or delay.

7.4 Satellite and Cellular Connectivity

As drones begin to operate over longer distances, especially in areas beyond the operator's line of sight (BVLOS), satellite and cellular connectivity are becoming increasingly essential components of drone communication systems. These technologies offer the potential for long-range, global connectivity, enabling drones to fly over remote regions, across cities, or even internationally.

Satellite Connectivity

Satellite communication (Satcom) provides a reliable means for drones to communicate over vast distances and in areas where terrestrial communication systems like Wi-Fi or cellular networks may not be available. Drones that are equipped with satellite communication systems can operate in remote or inaccessible regions, such as deep forests, deserts, or the open sea.

Satellite connectivity is achieved through the use of onboard satellite communication systems, such as small antennas and transceivers, which allow drones to connect to satellites orbiting the Earth. This system ensures that even if the drone is flying outside of line-of-sight or beyond cellular network range, it can still receive commands from the ground control station and transmit real-time data back.

The use of satellite communication in drones enables global operations, as it works anywhere on the planet. However, it also comes with some challenges, such as the need for expensive satellite transceivers, high data transmission costs, and the potential for signal interference caused by weather conditions or the drone's altitude.

Cellular Connectivity

The rise of 4G LTE and 5G networks has brought cellular communication into the forefront of drone operations, particularly for long-range missions that require real-time data transmission. Cellular connectivity allows drones to operate beyond the range of traditional RF communication systems and makes it possible to maintain continuous communication with ground control stations even when the drone is far away.

4G and 5G technologies offer several advantages over traditional RF-based communication systems. These networks provide high bandwidth, allowing drones to transmit large amounts of data, such as HD video streams and real-time telemetry, over long distances without the need for specialized equipment like satellite communication. Furthermore, cellular networks offer low-latency communication, which is crucial for real-time control and data transmission.

With the advent of 5G, drones can expect even faster data transmission speeds, lower latency, and greater capacity, making it ideal for applications such as autonomous delivery, smart city operations, and large-scale infrastructure monitoring.

7.5 Security and Privacy in Communication

As drones become more integrated into various industries, the need for robust security and privacy measures in communication systems has become increasingly critical. Drones transmit sensitive data, including location information, video feeds, and sensor data, all of which must be protected from unauthorized access, interception, and manipulation.

Securing Communication Channels

The security of drone communication systems is essential to ensure that data remains confidential and that the drone operates safely and securely. One of the most common threats to drone communication systems is **signal jamming** or **interference**, where malicious actors disrupt the signal between the drone and its ground control station. To mitigate this, drones must be equipped with encryption protocols to secure the transmission of data, making it difficult for attackers to intercept or tamper with the signals.

Common encryption methods used in drone communication include **Advanced Encryption Standard (AES)**, which is widely regarded as secure and is often used to protect RF communication channels. Additionally, **TLS (Transport Layer Security)** is commonly employed for securing data transmission over the internet, particularly when drones rely on cellular or satellite communication for long-range operations.

Privacy Concerns

Drones equipped with cameras and sensors often collect vast amounts of data that could potentially infringe on privacy rights. In applications such as surveillance, law enforcement, or media production, drones can capture sensitive information about individuals, locations, or activities. It is essential to address privacy concerns by ensuring that data collected by drones is protected and used ethically.

Governments and regulatory bodies have implemented various measures to address these concerns, such as defining no-fly zones around sensitive areas (e.g., private properties, government buildings) and ensuring that drones are used in compliance with privacy laws. Additionally, anonymization techniques can be applied to video feeds and sensor data to protect the identities and privacy of individuals.

In conclusion, ensuring secure communication systems and safeguarding privacy are paramount for the continued development and widespread adoption of drone technology. By integrating advanced encryption protocols, secure transmission methods, and privacy-aware designs, drones can be used safely and responsibly in a wide range of applications.

Chapter 8: Applications in Agriculture

8.1 Precision Farming and Crop Monitoring

In recent years, drones have revolutionized agricultural practices, particularly in the field of precision farming. Precision farming refers to the use of advanced technologies to monitor, manage, and optimize the growth of crops. Drones, with their ability to capture high-resolution imagery and collect data in real-time, play a crucial role in enhancing the efficiency and productivity of agricultural operations.

At the heart of precision farming is the use of data to make informed decisions about crop management. Drones equipped with various sensors, including multispectral, thermal, and RGB cameras, are able to fly over fields and capture detailed images of crops. These images provide farmers with valuable insights into the health of their crops, which can be used to make targeted interventions, such as adjusting irrigation or applying fertilizers more precisely.

One of the main applications of drone technology in precision farming is crop monitoring. By using drones to capture high-resolution images, farmers can monitor crop health, identify stress signs, and track growth stages. Drones provide an accurate, up-to-date view of the field, which is often more reliable than traditional scouting methods that rely on human observation. The ability to detect variations in plant health is essential for making data-driven decisions. For example, drones can help detect issues like nutrient deficiencies, diseases, or pest infestations, allowing farmers to take corrective measures before these problems spread.

Multispectral and hyperspectral imaging is particularly useful in crop monitoring, as these sensors can capture wavelengths of light beyond what the human eye can perceive. This allows the detection of crop stress before it becomes visible to the naked eye, giving farmers a head start in managing potential problems. For example, areas with poor vegetation health may be identified by capturing the differences in reflectance between healthy and stressed crops. This data is then analyzed using specialized software to generate vegetation indices like the Normalized Difference Vegetation Index (NDVI), which highlights areas of the field that require attention.

In addition to monitoring the health of crops, drones can also be used to assess other factors that affect crop growth, such as soil moisture levels and temperature. Drones equipped with thermal sensors can capture temperature data, which is useful for detecting variations in moisture levels. This data allows farmers to optimize irrigation practices, ensuring that crops receive the right amount of water at the right time. By preventing over- or under-watering, drones help conserve water resources, which is particularly important in areas facing water scarcity.

Furthermore, drones contribute to yield estimation by providing accurate data about crop density and overall plant health. With the help of high-resolution imagery and advanced data analytics, drones can predict how much yield a field will produce. This information helps farmers plan harvests more effectively and optimize resource allocation.

Overall, precision farming and crop monitoring powered by drones offer significant advantages, including increased efficiency, reduced input costs, improved crop yields, and sustainability. By enabling real-time monitoring and data-driven decision-making, drones are transforming the way farmers manage their crops.

8.2 Pest Control and Spraying Systems

Pest control is a major concern for farmers worldwide, as pests and diseases can devastate crops, leading to significant economic losses. Traditionally, pest control methods involved spraying pesticides across entire fields, often leading to overuse of chemicals, environmental harm, and increased operational costs. Drones, however, are providing farmers with more efficient and targeted solutions for pest control.

Drones equipped with spraying systems are capable of applying pesticides, herbicides, and fungicides directly to the affected areas of a field, minimizing the overall chemical usage. These systems often consist of liquid tanks, nozzles, and a flight control system that ensures precise delivery of the chemicals. By using GPS technology and advanced mapping software, drones can follow pre-programmed flight paths, ensuring that pesticides are applied only where they are needed, rather than uniformly across the entire field.

This targeted approach helps reduce the overall use of chemicals, leading to cost savings for farmers and a reduction in environmental impact. Over-application of chemicals can lead to runoff, contamination of water sources, and harm to beneficial insects, such as pollinators. By using drones to spray pesticides only in areas where pest pressure is high, farmers can reduce these risks while maintaining effective pest control.

Drones also play an important role in monitoring pest activity. By capturing high-resolution imagery of fields, drones can help farmers identify areas where pests are most active. For example, drones can detect changes in plant color or damage caused by pests, allowing farmers to quickly respond to emerging pest issues. The ability to detect pests early on can lead to more targeted pest control measures, which are both more effective and environmentally friendly.

In addition to pest control, drones are used for applying fertilizers and nutrients. The ability to apply fertilizers precisely and efficiently not only improves crop health but also reduces nutrient runoff and minimizes the impact on surrounding ecosystems. Using drones for precision spraying enables farmers to optimize their inputs, which in turn improves the sustainability and profitability of their operations.

While drone-based spraying systems offer numerous advantages, there are still challenges to be addressed. Regulatory issues surrounding the use of drones for pesticide application, including certification and licensing requirements, must be carefully managed. Additionally, the technology behind drone spraying systems needs to continue to evolve to handle larger fields and more complex crop systems effectively. However, the potential for drones to revolutionize pest control and spraying in agriculture is clear.

8.3 Soil and Field Analysis

Soil health is a critical factor in determining crop productivity, and understanding soil conditions is essential for making informed decisions about crop management. Traditionally, soil analysis involved manual sampling and laboratory testing, which can be time-consuming and costly. Drones, however, have introduced a more efficient and scalable way to analyze soil health and field conditions.

Drones equipped with multispectral sensors and other imaging technologies are capable of providing detailed information about soil moisture, texture, and nutrient content. These sensors capture data that is used to generate soil maps, which are invaluable for managing fertilizer applications, irrigation, and other aspects of crop care. Soil moisture maps, for example, can identify areas of a field that require additional irrigation or where water might be wasted, enabling farmers to optimize water usage.

Additionally, drones can be used to assess field variability. By flying over fields and capturing high-resolution images, drones can detect variations in soil conditions, such as areas with poor drainage or compacted soil. These variations can impact crop growth and yield, and addressing them early can help optimize the overall productivity of a field. Drones can also help farmers identify zones with different soil types, allowing for precision agriculture practices, such as variable rate fertilizer application, that cater to the unique needs of each zone.

Thermal sensors on drones can also play a key role in analyzing soil and field conditions. By capturing temperature data, drones can identify areas where the soil is warmer or cooler than the surrounding areas. Temperature variations in soil can indicate differences in moisture levels or organic matter content, helping farmers to target their management practices more effectively.

In addition to soil health, drones can be used to monitor crop stress, which is often linked to soil conditions. By analyzing the imagery captured by drones, farmers can identify areas where crops are underperforming due to soil-related issues. This early detection of stress allows farmers to intervene before crop yield is significantly affected, leading to higher productivity and better resource management.

Overall, drones offer an effective and efficient means for soil and field analysis. They provide farmers with real-time, actionable insights that can help optimize crop production, reduce waste, and improve overall farm sustainability.

8.4 Livestock Management

Livestock management is an essential aspect of modern agriculture, and drones are proving to be valuable tools in this field. From monitoring herd health to tracking animals in large pastures, drones are helping farmers manage their livestock more effectively, improving productivity and animal welfare.

Drones equipped with cameras and thermal imaging sensors are used to monitor livestock health, detect illnesses, and ensure that animals are properly cared for. For example, drones can capture real-time images of livestock, which can then be analyzed to detect signs of injury or disease. By using thermal cameras, drones can identify animals that are running a fever or exhibiting abnormal behavior, which are often signs of illness. Early detection allows farmers to isolate affected animals and take appropriate action, such as administering treatments, reducing the spread of disease, or adjusting feeding schedules.

Drones are also used to track the movement of livestock across large areas, which is especially useful in extensive grazing systems where animals roam freely across large pastures. Using GPS and thermal imaging, drones can locate individual animals or groups of animals, ensuring that they remain within designated areas and are not lost or separated from the herd. This capability is particularly valuable for farmers who manage large numbers of livestock in vast, remote areas, where it may be difficult to physically monitor all animals.

Moreover, drones can assist with herd management by helping farmers monitor grazing patterns. By using drones to observe how animals are grazing across pastures, farmers can determine whether grazing pressure is balanced or if certain areas are being overgrazed. Overgrazing can lead to land degradation, reduced pasture productivity, and loss of biodiversity. By closely monitoring grazing patterns, drones enable farmers to implement better grazing management strategies, ensuring the long-term health of both the livestock and the land.

In addition, drones are used for more routine tasks, such as checking fences and water sources. By flying over pastures, drones can inspect the condition of fences and other infrastructure, reducing the need for manual inspections. This not only saves time but also helps prevent potential problems before they become critical.

Overall, drones are enhancing livestock management by improving animal health monitoring, optimizing grazing practices, and providing valuable data for better decision-making.

8.5 Benefits and Challenges in Agri-Tech Adoption

The adoption of drones in agriculture has brought numerous benefits to the sector, but it has also presented challenges that must be addressed for successful integration. As the agricultural industry continues to embrace drone technology, it is important to examine both the benefits and the challenges that come with it.

Benefits

One of the primary benefits of using drones in agriculture is the ability to collect vast amounts of data in real-time. This data allows farmers to make informed decisions, leading to increased efficiency, reduced costs, and improved productivity. Drones enable precision agriculture by allowing farmers to apply inputs like water, fertilizers, and pesticides only where they are needed, minimizing waste and reducing environmental impact.

Another benefit is the ability to monitor crop health and soil conditions more effectively. Drones provide farmers with detailed, up-to-date images of their fields, allowing them to identify problems early and take corrective action before they become major issues. This improves overall crop yields and reduces the need for costly interventions.

Drones also improve efficiency by reducing the need for manual labor in tasks like crop monitoring, pest control, and field analysis. This allows farmers to focus on more strategic aspects of their operations while automating routine tasks.

Challenges

Despite the many advantages, there are challenges to the widespread adoption of drone technology in agriculture. One of the major challenges is the high initial cost of drone equipment, including the purchase of drones, sensors, and software. Although the long-term benefits can outweigh the costs, the upfront investment may be prohibitive for small-scale farmers.

Regulatory hurdles also pose a significant challenge to the widespread use of drones in agriculture. Many countries have strict regulations governing the use of drones, including requirements for pilot certification, flight restrictions, and no-fly zones. Navigating these regulations can be time-consuming and complex, particularly for farmers who may not have experience with drone operations.

Moreover, the integration of drones into existing agricultural workflows can be challenging. While drones provide valuable data, farmers must also have the knowledge and tools to interpret and act on that data. The need for specialized software, training, and data analysis skills can make it difficult for some farmers to adopt drone technology without significant investment in education and infrastructure.

Finally, there are concerns regarding the security and privacy of drone-collected data. As drones capture high-resolution images and sensor data, there is a risk that this information could be accessed or misused by unauthorized parties. Ensuring the privacy and security of agricultural data is essential to fostering trust in drone technology.

In conclusion, drones offer numerous benefits to the agriculture sector, including enhanced precision farming, improved pest control, better soil analysis, and more efficient livestock management. However, challenges such as high costs, regulatory issues, and the need for specialized skills must be addressed to fully realize the potential of drone technology in agriculture. With continued innovation and support, drones are set to play a transformative role in the future of agriculture.

Chapter 9: Applications in Delivery and Logistics

9.1 Drone-Based Package Delivery Systems

The integration of drones into delivery and logistics systems has been one of the most anticipated applications of UAV technology. Drone-based package delivery systems are designed to transport goods quickly, efficiently, and with minimal human intervention. This application has the potential to revolutionize industries ranging from e-commerce to healthcare, by reducing delivery times, costs, and improving customer satisfaction.

The concept of drone-based package delivery emerged as an answer to the challenges posed by traditional delivery methods. With the rise of e-commerce, there is an increasing demand for faster delivery solutions. Drones, with their ability to navigate directly from the warehouse to the delivery location without the need for roads or traffic, can significantly shorten delivery times. This is especially true for high-priority or time-sensitive deliveries, such as medical supplies or small consumer goods.

Drone-based package delivery systems rely on a combination of technologies, including GPS navigation, automated flight control, and robust communication networks. Typically, a customer places an order online, and the package is prepared for dispatch from a distribution center. Once the package is secured in the drone, the system calculates the most efficient route based on distance, weather conditions, and other factors, ensuring the fastest and safest delivery.

One of the primary advantages of using drones for package delivery is the reduction in human labor and transportation costs. Drones eliminate the need for delivery trucks and drivers, significantly reducing the cost per delivery, particularly for short distances or remote areas. Additionally, the ability to carry small, lightweight packages allows drones to make deliveries much faster than traditional methods. For instance, drones can deliver goods in minutes, while conventional delivery methods could take hours or even days.

Many companies are already experimenting with and deploying drone-based delivery systems. In 2016, Amazon introduced its Prime Air service, which aims to deliver packages weighing up to five pounds within 30 minutes or less using drones. Similarly, Google's Project Wing has developed drone delivery systems for remote areas, with a focus on disaster relief and medical deliveries.

However, drone-based package delivery is not without its challenges. One of the primary concerns is payload capacity, as current drones are typically limited to small, lightweight packages. Most commercial drones can carry between 2 to 5 kilograms of weight, which may not be sufficient for larger packages. Furthermore, there are technological and regulatory hurdles to overcome, such as developing secure, reliable navigation systems that ensure drones avoid obstacles and deliver packages to precise locations. The systems also need to be scalable, able to manage large numbers of deliveries simultaneously, and capable of maintaining consistent delivery quality.

Drone-based package delivery systems hold great promise for the future of logistics. However, continued innovation in drone design, autonomous flight technology, and regulatory frameworks will be crucial for the widespread adoption and success of these systems.

9.2 Last-Mile Delivery Solutions

The "last mile" of delivery is often considered the most challenging and expensive part of the logistics chain. This phase involves transporting goods from a central distribution center to the final destination—typically, the customer's doorstep. In urban areas, this can involve navigating congested streets, dealing with traffic, and managing parking challenges. In rural or remote areas, the last mile can be even more difficult, requiring long travel distances on poorly maintained roads. Drones offer a promising solution to these problems by providing a more direct and efficient method of delivery.

Last-mile delivery refers to the final leg of a product's journey from the distribution center to the customer. It is often characterized by high costs due to the complexity of reaching individual consumers, especially in densely populated cities or rural areas. Traditional methods of last-mile delivery, such as trucks or vans, rely on road infrastructure and are subject to delays caused by traffic congestion, roadworks, and weather conditions.

Drones, however, can fly directly to a customer's location, bypassing the need for roads entirely. With their ability to avoid traffic, drones can significantly reduce delivery times, especially in urban areas with heavy congestion. Moreover, drones are able to access hard-to-reach locations, such as remote rural areas, mountainous terrains, or islands, where road infrastructure is lacking or less developed.

The process of last-mile drone delivery begins with the dispatch of a drone carrying the package from a central warehouse or delivery hub. GPS systems and advanced flight planning software guide the drone to the customer's address. Drones are typically equipped with secure, weather-resistant cargo holds to protect the packages during transport. Once the drone reaches its destination, the package is either dropped in a designated area or delivered to a specified location, such as a porch or backyard.

One of the most significant advantages of drone-based last-mile delivery is speed. Traditional delivery services often take hours or even days to complete the final leg of a delivery, whereas drones can potentially complete this process in a matter of minutes. The use of drones reduces the need for delivery trucks, thus saving costs associated with fuel, maintenance, and labor. Furthermore, drones can reduce the environmental footprint of delivery operations, as they are often electric-powered, emitting fewer pollutants than traditional vehicles.

Despite the potential advantages, last-mile drone delivery faces several challenges. One of the main obstacles is the regulatory landscape, as many countries have strict rules governing the use of drones for commercial purposes. These include restrictions on flight paths, no-fly zones, and limitations on the maximum weight a drone can carry. Furthermore, drones must be able to safely navigate urban environments, avoiding obstacles such as buildings, power lines, and trees.

The technology behind last-mile delivery drones is advancing rapidly, with innovations such as autonomous flight systems, enhanced sensors, and artificial intelligence (AI) playing a key role in making these systems safer and more reliable. However, widespread adoption of drone-based last-mile delivery will depend on resolving regulatory, technical, and logistical challenges.

9.3 Inventory Management in Warehouses

Inventory management is a critical function in logistics, involving the tracking, storage, and retrieval of goods in warehouses and distribution centers. Efficient inventory management ensures that businesses can fulfill orders promptly, minimize stockouts, and maintain accurate records. In recent years, drones have emerged as a valuable tool in optimizing inventory management processes, improving efficiency, and reducing operational costs.

In a traditional warehouse, inventory management relies on human workers to manually scan barcodes, move items, and take stock. This process can be time-consuming and prone to errors, especially in large warehouses with complex inventories. Drones, however, offer a more efficient and automated solution to these challenges.

Equipped with RFID (Radio Frequency Identification) or barcode scanning capabilities, drones can autonomously fly through warehouse aisles, identifying and recording the location of products. These drones are programmed to follow pre-determined flight paths, using advanced navigation systems to avoid obstacles and ensure they are operating within the warehouse's layout. Drones can autonomously scan labels on products, cross-reference them with inventory databases, and update stock levels in real-time.

The use of drones for inventory management provides several benefits. First, drones increase the speed and accuracy of inventory counts. By automating the scanning process, drones can complete tasks much faster than human workers, reducing the time required for stocktaking. Additionally, drones eliminate the risk of human error, ensuring that inventory data is more accurate. This leads to better decision-making, as warehouse managers have access to real-time, reliable data about stock levels and locations.

Another significant advantage is cost savings. With drones handling inventory management tasks, businesses can reduce the number of workers needed for these tasks, thereby cutting labor costs. Furthermore, drones are capable of operating in challenging environments, such as high shelves or narrow aisles, which may be difficult for human workers to access.

Despite the advantages, several challenges remain in the implementation of drones for warehouse inventory management. One of the key challenges is integrating drone systems with existing warehouse management software. For drones to be fully effective, they need to work seamlessly with the software systems that track inventory levels, order fulfillment, and shipment logistics. Furthermore, there are technical limitations, such as battery life and payload capacity, that can restrict the range and functionality of drones in large warehouses.

Overall, drones offer significant improvements in inventory management by enhancing speed, accuracy, and cost-efficiency. As the technology evolves, drones are expected to play a growing role in warehouse operations.

9.4 Disaster Relief and Medical Supply Distribution

Drones have emerged as a powerful tool for disaster relief and medical supply distribution, particularly in areas that are difficult to access due to infrastructure damage, natural disasters, or geographical challenges. In emergency situations, timely delivery of medical supplies, food, water, and other critical resources is essential for saving lives and supporting recovery efforts. Drones, with their ability to quickly navigate over damaged roads and difficult terrain, offer an efficient and effective solution for delivering these supplies to the affected areas.

During natural disasters such as earthquakes, floods, or hurricanes, traditional delivery methods may be hindered by damaged infrastructure, such as collapsed bridges or blocked roads. Drones can bypass these obstacles, flying over flooded areas, landslides, or debris to deliver supplies directly to areas in need. In addition, drones can provide real-time imagery to emergency responders, helping them assess damage and plan rescue operations more effectively.

One of the most notable applications of drones in disaster relief is the delivery of medical supplies. Drones are already being used to deliver vaccines, blood products, insulin, and other medical essentials to remote or underserved areas. For example, in Rwanda, drones are used to deliver blood and medical supplies to rural clinics, cutting down delivery times from several hours to just minutes. Similarly, in places like Haiti, drones have been used to deliver medicines and emergency kits following natural disasters.

In addition to delivering medical supplies, drones are also used for aerial surveys and mapping. Drones equipped with high-resolution cameras and thermal imaging sensors can provide detailed images of disaster zones, helping responders assess the extent of the damage and identify areas that require immediate attention. This data is invaluable for coordinating relief efforts and ensuring that resources are directed where they are most needed.

Despite their potential, the use of drones in disaster relief faces several challenges. One of the major issues is regulatory approval, as many countries have strict regulations governing the use of drones, particularly in emergency situations. Additionally, drones require a reliable communication infrastructure to function effectively, which may be disrupted during disasters. Weather conditions, such as high winds or heavy rain, can also limit the ability of drones to operate safely and reliably.

Overall, drones have proven to be invaluable tools for disaster relief and medical supply distribution. Their ability to deliver critical supplies to hard-to-reach areas and provide real-time data is transforming emergency response operations.

9.5 Regulatory Hurdles and Public Perception

The rapid development and deployment of drone technology in delivery and logistics has raised several regulatory and public perception issues. Governments around the world are grappling with how to regulate the use of drones, particularly for commercial applications like package delivery. Public concerns about safety, privacy, and the potential impact of drones on jobs have added complexity to the regulatory landscape.

One of the key challenges is airspace management. In many countries, airspace is highly regulated, with specific zones designated for commercial, military, and recreational aircraft. The integration of drones into this airspace requires careful coordination to ensure that they do not interfere with manned aircraft or create safety hazards. Drone operators must adhere to rules regarding altitude limits, no-fly zones, and flight paths, which vary depending on the country and region.

Another challenge is ensuring that drones are operated safely, particularly in urban areas where there are higher risks of collisions with buildings, vehicles, or people. As a result, many countries require drone operators to obtain licenses or certifications and to comply with strict safety standards. Some countries have implemented mandatory drone insurance policies to protect against accidents and damage caused by drone operations.

Public perception of drones is another significant factor that can influence the adoption of drone-based delivery systems. While drones offer numerous benefits in terms of efficiency and convenience, there are concerns about their impact on privacy. Drones are capable of capturing high-resolution imagery and sensor data, which could be used to monitor individuals or collect sensitive information without their consent. Addressing these concerns through proper data privacy regulations and transparent drone operations is crucial for gaining public trust.

Additionally, there are concerns about job displacement due to the automation of delivery systems. As drones become more widely used for package delivery, some fear that traditional delivery jobs, such as truck drivers and couriers, may be replaced. While drones may reduce the need for certain manual labor tasks, they also create new job opportunities in fields like drone maintenance, data analysis, and system management. Ensuring a balanced approach that supports workers transitioning to new roles is essential for mitigating these concerns.

In conclusion, while drones offer tremendous potential for transforming the delivery and logistics industries, regulatory hurdles and public concerns must be carefully addressed. By developing clear regulations and fostering open communication with the public, the full benefits of drone technology can be realized in delivery systems.

Chapter 10: Drones in Surveillance and Security

10.1 Role in Law Enforcement

Drones have rapidly become an essential tool in modern law enforcement due to their ability to enhance situational awareness, improve operational efficiency, and reduce risk to officers. As part of the technological evolution in policing, drones provide law enforcement agencies with advanced capabilities for monitoring, tracking, and responding to incidents. Their potential in surveillance, crowd control, and even search-and-rescue operations has made them an indispensable asset in modern policing strategies.

One of the most prominent roles that drones play in law enforcement is providing aerial surveillance. With their ability to fly at high altitudes and cover large areas, drones can capture real-time imagery or video, enabling officers to monitor criminal activity from a safe distance. For instance, drones are used during high-speed chases to provide aerial support, giving officers an overview of the situation and helping them make informed decisions regarding road closures or coordination with ground units.

Drones are also valuable in monitoring public spaces, which is especially critical in situations like protests or large gatherings. By using drones for aerial reconnaissance, law enforcement agencies can assess the size, movement, and potential risks of crowds, all without physically engaging with the crowd. This not only minimizes potential confrontations but also provides valuable intelligence that can guide law enforcement decisions.

Another significant application of drones in law enforcement is during search-and-rescue missions. When someone goes missing in a difficult-to-reach area, such as a forest, mountains, or even an urban environment after a disaster, drones equipped with thermal imaging cameras can locate individuals quickly. This capability allows officers to narrow down the search area, reducing the time and manpower needed to locate and rescue individuals.

Furthermore, drones are equipped with specialized sensors and cameras, such as infrared and thermal imaging, which can detect heat signatures, even in low-light or no-light conditions. This makes drones particularly effective for finding suspects or missing persons, even in the cover of darkness.

However, the increased use of drones in law enforcement has raised concerns regarding privacy and civil liberties. The surveillance capabilities of drones could potentially infringe on individuals' right to privacy if not properly regulated. There is a fine balance that law enforcement agencies must strike between using drones for legitimate public safety purposes and respecting individuals' privacy rights.

In many countries, laws governing the use of drones by law enforcement are being developed to ensure that their use is consistent with privacy protections. This may include limitations on the types of surveillance that can be conducted, who can authorize drone use, and how data gathered by drones is stored and processed. These regulations are critical for ensuring that drones are used responsibly and ethically in law enforcement operations.

10.2 Border Patrol and Coastal Monitoring

The use of drones for border patrol and coastal monitoring has gained significant traction as governments seek more efficient ways to secure borders and monitor coastlines. The vast, often remote and challenging terrain of borders and coastal areas makes traditional methods of surveillance difficult and costly. Drones offer an innovative solution by providing real-time intelligence over large areas without the need for a heavy ground presence or the limitations of manned aircraft.

In border security, drones are used extensively to patrol large stretches of land that are difficult to monitor with traditional means. This is especially important for countries with long and porous borders, where illegal activities such as drug trafficking, human smuggling, and illegal immigration are common concerns. Drones can fly at high altitudes, covering expansive areas quickly, while simultaneously monitoring potential entry points and detecting any unauthorized crossings. By using high-resolution cameras, infrared sensors, and even radar, drones can identify individuals or vehicles attempting to cross borders illegally and alert border patrol agents to their location.

In coastal areas, drones are used to monitor both the shoreline and the waters surrounding the coast. Drones provide a significant advantage over traditional boat patrols or stationary surveillance equipment by offering flexibility in how and where monitoring occurs. They can patrol large distances of coastline, checking for signs of illegal fishing, human trafficking, or even environmental pollution. In areas prone to natural disasters like hurricanes or tsunamis, drones can assist in monitoring the condition of coastal infrastructure, assess the extent of damage, and provide real-time information for rescue operations.

The capability of drones to operate in adverse weather conditions further enhances their effectiveness in both border patrol and coastal monitoring. With the integration of advanced weather-resistant technologies and specialized cameras, drones can operate in harsh environments such as extreme heat, cold, or rain. They are also able to fly at low altitudes, enabling them to monitor remote or difficult-to-reach locations like mountains, caves, or islands along the border or coastline.

Drones used for border and coastal monitoring are often equipped with sensors and imaging systems such as thermal cameras, GPS tracking, radar, and motion detectors. This equipment allows them to identify targets based on heat signatures or movement patterns, even in the dark. Drones can also send real-time data back to operators, allowing for immediate responses to suspicious activities or security breaches.

The use of drones in border patrol and coastal monitoring has proven to be a cost-effective solution compared to the use of manned patrols or stationary surveillance systems. The ability to cover vast stretches of land and sea without the need for large teams or expensive infrastructure provides significant savings. Drones can operate autonomously, reducing the need for human intervention and minimizing risks to personnel.

Despite these benefits, the deployment of drones for border and coastal surveillance raises several ethical and legal concerns. There are questions about the extent to which drones can be used to monitor individuals without violating privacy rights. Some critics argue that the increased use of drones in border security could lead to excessive surveillance, potentially infringing on citizens' freedoms. Regulatory frameworks need to be in place to ensure drones are used appropriately, with proper oversight to avoid misuse of the technology.

10.3 Public Safety During Events and Emergencies

Drones have proven to be invaluable tools for ensuring public safety during events and emergencies, where large crowds, time-sensitive situations, and complex logistics present significant challenges. From monitoring public events to coordinating responses during emergencies, drones enhance situational awareness, increase response efficiency, and improve overall safety.

During large public events, such as concerts, sporting events, or rallies, drones provide a bird's-eye view of the crowd, helping law enforcement and event organizers monitor crowd behavior and identify potential risks. Drones can be equipped with high-definition cameras, thermal imaging, and facial recognition software to track individuals and detect abnormal behavior or threats. For example, drones can identify individuals acting suspiciously or engaging in criminal activity, enabling security personnel to respond quickly and prevent incidents from escalating.

Drones also enhance public safety by improving the coordination of emergency responders in real-time. For instance, during fires, accidents, or terrorist attacks, drones can provide critical aerial footage that helps emergency services assess the scene, determine the best course of action, and guide their response teams. Drones can fly into dangerous areas, such as burning buildings or hazardous materials zones, to capture real-time data, which can be relayed back to first responders, thus informing their decisions and ensuring that they operate more safely and effectively.

In the case of search and rescue operations, drones can be deployed to quickly cover vast areas and locate missing persons. Equipped with thermal imaging cameras, drones can find individuals who are lost, trapped, or stranded in remote areas, even in low-visibility conditions like thick forests or darkened environments. This technology significantly reduces the time required to locate missing individuals, which can be critical in saving lives during emergencies.

Additionally, drones are increasingly being used in response to natural disasters. When traditional infrastructure is damaged or inaccessible, drones can provide real-time imagery of disaster zones, allowing response teams to plan their actions more effectively. Drones can also transport small supplies, such as medical kits or communication devices, to hard-to-reach areas, helping people stranded or in urgent need of assistance.

While drones offer substantial benefits in public safety, their widespread use during events and emergencies also brings potential risks. Concerns over privacy, safety, and data security are at the forefront of discussions surrounding drones in public spaces. Additionally, issues such as drone malfunctions, no-fly zones, and potential interference with other aircraft must be addressed to ensure drones can operate effectively without causing harm or confusion.

10.4 Anti-Drone Systems and Countermeasures

As drones have become more prevalent in surveillance and security operations, there has been an increasing focus on developing anti-drone systems and countermeasures. While drones offer significant advantages, they also pose a threat in certain contexts, such as when they are used for criminal activities, unauthorized surveillance, or even as weapons in attacks. In response to these threats, numerous counter-drone technologies have been developed to neutralize or disable drones in a secure, controlled manner.

Anti-drone systems typically rely on various technologies, including jamming, spoofing, and interception methods. Jamming works by emitting high-frequency signals that interfere with the communication between a drone and its operator, effectively disabling the drone's navigation system. Spoofing, on the other hand, involves sending false signals to the drone, causing it to believe it is in a different location or leading it to return to its starting point. Some counter-drone systems also utilize directed energy, such as lasers, to physically disable drones by targeting their electronic systems or causing them to overheat and crash.

A more physical approach to countering drones involves intercepting or capturing them. This can be done using nets, either launched by other drones or fired from ground-based systems, to ensnare the target drone and bring it down safely. Some systems deploy trained birds of prey, such as eagles, to physically snatch drones out of the air. While these methods are less common, they have demonstrated success in certain scenarios, particularly in protecting sensitive areas like airports or critical infrastructure.

Despite the success of various counter-drone systems, there are significant challenges associated with their development and deployment. One of the main issues is that many anti-drone systems can be expensive and require specialized personnel to operate. Additionally, the use of countermeasures must be carefully regulated to avoid unintended consequences, such as interference with legitimate drone operations or other communication systems.

In many countries, laws surrounding the use of counter-drone technology are still in development. Public concerns about safety and the potential for misuse of these systems add complexity to the regulatory landscape. However, as drone-related threats continue to grow, governments and private entities will likely invest more in developing effective countermeasures to secure sensitive areas and prevent malicious drone activity.

10.5 Ethical Implications of Surveillance

The use of drones in surveillance raises significant ethical concerns that need to be carefully considered. Drones, with their ability to monitor large areas and gather detailed data, have the potential to infringe on privacy rights, especially when used in public spaces without consent. These concerns are amplified by the growing sophistication of drone technology, including facial recognition software, thermal imaging, and data collection capabilities that can track individuals and monitor their activities in real-time.

One of the most pressing ethical questions is the extent to which drones should be allowed to surveil individuals in public or private spaces. While drones can provide valuable intelligence for law enforcement and security purposes, their use must be balanced with the right to privacy. Many argue that unrestricted surveillance by drones could lead to a "Big Brother" society, where individuals are constantly monitored without their knowledge or consent.

Privacy concerns are compounded by the ease with which drones can gather and store data. With high-definition cameras and advanced sensors, drones can capture images, videos, and other data from vast distances and transmit this information back to operators. In some cases, drones may even be able to collect data that individuals are not aware they are producing, such as thermal signatures or audio recordings.

Another ethical consideration is the use of drones for crowd surveillance and monitoring public events. While drones can help identify criminal activity or prevent violence, they also raise concerns about the potential misuse of surveillance technologies. The ability to track individuals in a crowd, for example, could be used for purposes beyond public safety, such as political or social control.

The ethical implications of drone surveillance are further complicated by the question of accountability. When drones are used in surveillance, it can be difficult to determine who is responsible for the data they collect, how that data is used, and whether it is properly safeguarded. There is a need for clear regulations and oversight to ensure that drones are used responsibly and that individuals' rights are protected.

In conclusion, while drones offer significant benefits for security and surveillance, their ethical implications cannot be ignored. Striking a balance between the need for public safety and the protection of individual freedoms is essential to ensuring that drone technology is used responsibly and ethically. Legal frameworks, transparency, and accountability will play critical roles in addressing these concerns.

Chapter 11: Environmental and Scientific Research

11.1 Monitoring Climate Change and Ecosystems

Drones have emerged as powerful tools for environmental monitoring, providing researchers with the ability to collect real-time data from remote, difficult-to-reach, or hazardous locations. One of the most critical areas where drones have been applied is in monitoring climate change and ecosystems, two phenomena that are intricately connected. Climate change is affecting ecosystems worldwide, with consequences such as shifting weather patterns, rising sea levels, and changes in biodiversity. Understanding and mitigating these changes requires detailed, accurate, and continuous monitoring, and drones play a pivotal role in enabling such surveillance.

Drones are equipped with a wide array of sensors that can monitor environmental variables, such as temperature, humidity, air quality, and greenhouse gas concentrations. Their ability to fly at varying altitudes and over large areas allows for the collection of comprehensive data on ecosystems, even in remote or dangerous regions. For example, drones equipped with infrared and multispectral cameras can provide valuable insights into vegetation health, forest density, and changes in land cover, all of which are critical for assessing the impacts of climate change. In tropical rainforests, which are among the most biodiverse ecosystems on Earth, drones can track deforestation rates, monitor the effects of logging, and identify areas where conservation efforts are needed.

One of the most effective uses of drones in climate change monitoring is their ability to track carbon sequestration in forests and wetlands. By mapping these areas and monitoring changes over time, drones provide valuable information on how these ecosystems are responding to climate change and how they can be better managed for carbon capture. Drones can also monitor melting glaciers, rising sea levels, and changes in ice cover, providing critical data for understanding global warming's effects on polar regions.

Additionally, drones are used to monitor other sensitive ecosystems such as coral reefs, wetlands, and mangrove forests. These ecosystems are particularly vulnerable to climate change, and their health is often an indicator of broader environmental trends. For example, drones equipped with high-resolution cameras and sensors can assess coral reef bleaching, a phenomenon linked to rising ocean temperatures. Similarly, drones can help track the loss of wetlands or mangrove areas due to rising sea levels and coastal development, providing critical data to inform conservation strategies.

In addition to their effectiveness in tracking the direct effects of climate change, drones also play a role in monitoring human activities that exacerbate environmental problems, such as illegal logging, poaching, and pollution. Drones are often deployed to track illegal activities in protected areas, providing real-time data that can assist authorities in taking swift action to protect sensitive ecosystems. Moreover, drones are often used to measure air pollution in urban areas, helping scientists understand the sources and impacts of pollution on both human health and the environment.

The ability of drones to gather data in a variety of formats, including high-definition video, multispectral images, and thermal data, has revolutionized environmental research. These diverse data types can be analyzed to assess ecosystem health, detect environmental stressors, and track changes over time. This continuous monitoring allows for the identification of trends and patterns that would otherwise be difficult to observe, enabling scientists to develop better models of climate change and its impacts.

However, the widespread use of drones in environmental monitoring is not without challenges. One of the primary concerns is data management. The vast amounts of data collected by drones can be overwhelming, and analyzing it efficiently requires advanced tools, such as artificial intelligence (AI) and machine learning, to process and interpret the information. Additionally, the accuracy and reliability of drone-collected data depend on the calibration of sensors and the quality of the flight paths, which require careful planning and execution.

Moreover, ethical concerns related to the use of drones in sensitive ecosystems must be addressed. The potential disturbance to wildlife and habitats, especially in fragile ecosystems, must be minimized, and proper guidelines should be followed to ensure that drones do not interfere with conservation efforts.

11.2 Wildlife Conservation Efforts

The use of drones in wildlife conservation has become increasingly popular due to their ability to monitor wildlife populations and habitats with minimal disruption. Drones are uniquely positioned to provide researchers with access to remote or difficult-to-reach areas, where traditional monitoring methods, such as ground surveys or manned aircraft, are often impractical or costly. In conservation efforts, drones help monitor both terrestrial and aquatic ecosystems, offering crucial insights into biodiversity, species behavior, and the health of endangered species.

In terrestrial ecosystems, drones are widely used to track animal populations, monitor migration patterns, and assess habitat conditions. For example, drones equipped with infrared or thermal imaging cameras can help locate animals in dense forests, grasslands, or even vast deserts, where ground-based surveys are labor-intensive and time-consuming. These drones are capable of flying at altitudes and distances that give them a comprehensive view of the landscape, enabling researchers to observe large areas and monitor animals without causing disturbances. This is especially important in the case of endangered species, such as elephants, rhinos, or big cats, where human presence may lead to stress or behavioral changes.

One of the key advantages of drones in wildlife conservation is their ability to monitor species over long periods. Researchers can use drones to conduct regular aerial surveys, allowing them to track the movement and health of animal populations in real time. This can be particularly useful in remote regions where other forms of monitoring, such as camera traps or manual tracking, are not feasible.

Drones are also instrumental in monitoring illegal activities that threaten wildlife, such as poaching, illegal logging, and habitat destruction. By providing real-time surveillance of protected areas, drones can assist park rangers in detecting and responding to criminal activities. For example, drones can be used to fly over national parks and wildlife reserves to detect poachers or track their movements, helping law enforcement agencies intercept them before they can cause harm to endangered species.

In aquatic ecosystems, drones are used to monitor marine life and protect fragile habitats such as coral reefs, seagrass beds, and marine sanctuaries. Equipped with high-definition cameras and sensors, drones can capture images and videos of underwater ecosystems, enabling scientists to study coral bleaching, fish populations, and changes in water quality. In some cases, drones are also used to monitor the movements of marine animals such as whales, sea turtles, and sharks, providing researchers with valuable data on their behavior and habitat preferences.

In addition to their ability to collect data, drones also play a significant role in wildlife conservation education and advocacy. Drones are often used to capture stunning aerial footage of wildlife and natural habitats, which can be shared with the public through social media, documentaries, or conservation campaigns. These images help raise awareness about the importance of conservation and the threats facing biodiversity, encouraging people to get involved and support conservation initiatives.

Despite their numerous advantages, the use of drones in wildlife conservation also presents challenges. One of the most significant concerns is the potential disturbance to animals. Even though drones are less intrusive than humans, they can still cause stress or alter the behavior of animals, particularly in sensitive species or habitats. To minimize these impacts, it is essential for conservation organizations to establish ethical guidelines for drone use, ensuring that drones are operated in a way that minimizes disturbance to wildlife and does not interfere with ongoing conservation efforts.

Another challenge is the regulation and oversight of drone usage in protected areas. As drones become more popular in wildlife conservation, the potential for misuse or accidental disruptions to wildlife habitats increases. Governments and conservation organizations must work together to develop and enforce regulations that ensure drones are used responsibly and effectively in conservation efforts.

11.3 Geological and Mineral Surveys

Drones have revolutionized the field of geological and mineral surveys, offering a cost-effective and efficient method of conducting fieldwork in areas that are otherwise difficult to access. Traditional geological surveys, particularly those in remote or hazardous environments, require significant manpower, equipment, and time. Drones, on the other hand, can rapidly survey large areas, provide high-resolution imagery, and collect valuable data without the need for extensive ground-based exploration.

One of the primary applications of drones in geology is mapping and topographic surveying. Equipped with LiDAR (Light Detection and Ranging) technology or high-resolution cameras, drones can create detailed three-dimensional maps of terrain, highlighting changes in elevation, soil composition, and geological features. This data is crucial for studying landforms, mapping faults and tectonic boundaries, and understanding the underlying structure of the Earth's surface.

Drones are particularly useful for monitoring and surveying areas that are prone to natural disasters, such as landslides, earthquakes, or volcanic eruptions. After a major geological event, drones can be deployed to rapidly assess the extent of damage and monitor any ongoing activity, such as aftershocks or lava flow. This enables geologists to gather real-time data, which is critical for understanding the dynamics of the event and for issuing timely warnings to communities in the affected areas.

In mineral exploration, drones have proven invaluable in helping identify areas that are rich in resources. By flying over mineral deposits and using remote sensing technologies, drones can detect valuable minerals, such as gold, copper, or lithium, as well as assess the condition of mining sites. Drones can also be used to inspect active mining sites, monitor the environmental impact of mining operations, and ensure compliance with regulations. In many cases, drones can replace or supplement traditional ground surveys, providing more accurate and less disruptive results.

Another key benefit of drones in geological surveys is their ability to provide continuous monitoring of sites over time. By conducting regular drone flights, researchers can track changes in geological features, monitor erosion patterns, and detect shifts in mineral deposits or the stability of structures such as cliffs or mines.

The challenges of using drones for geological surveys mainly involve the accuracy of the data collected. Factors such as weather conditions, terrain complexity, and the quality of the drone's sensors can affect the precision of the results. Additionally, in areas where regulations are not well-defined, drones may face restrictions on where they can operate, limiting their ability to fully serve the needs of geological researchers.

11.4 Atmospheric and Weather Data Collection

Drones have become increasingly important in atmospheric and weather research, providing scientists with the ability to collect detailed, real-time data from the Earth's atmosphere. Weather balloons and manned aircraft have traditionally been used to gather atmospheric data, but drones offer several advantages, including the ability to fly at lower altitudes, cover specific regions of interest, and operate in conditions that might be too dangerous or costly for other forms of data collection.

Drones equipped with atmospheric sensors can measure a wide range of variables, including temperature, humidity, barometric pressure, wind speed, and air quality. By flying at different altitudes and locations, drones can create vertical profiles of the atmosphere, offering insights into weather patterns and phenomena such as storms, tornadoes, and hurricanes. This real-time data collection can help meteorologists predict weather events more accurately and issue timely warnings to the public.

Drones also provide valuable data on air pollution, which is a growing concern in urban areas around the world. By flying over cities or industrial zones, drones can measure pollutant concentrations, such as nitrogen dioxide, sulfur dioxide, and particulate matter, helping to track sources of pollution and assess its impact on human health. This data is essential for developing effective environmental policies and addressing the challenges of climate change.

Furthermore, drones play an important role in studying extreme weather events. For example, drones have been used to gather data during hurricanes, providing insights into wind speeds, pressure changes, and storm dynamics. Drones are particularly useful for collecting data in hard-to-reach areas, such as the eye of a storm, where traditional methods of data collection are difficult or dangerous.

The integration of drones with weather forecasting models is another area where they have shown promise. By continuously collecting data and feeding it into predictive models, drones can enhance the accuracy and timeliness of weather predictions, improving disaster preparedness and response.

However, challenges related to the use of drones for atmospheric research include the potential interference with other aviation systems and the limitations of drone battery life. Operating drones in harsh weather conditions, such as thunderstorms or high winds, is also a significant challenge. Nonetheless, the growing capabilities of drones and their integration with other atmospheric technologies make them an invaluable tool for advancing weather research and improving our understanding of the Earth's atmosphere.

11.5 Role in Marine and Oceanic Research

Drones have also made a significant impact on marine and oceanic research, offering a cost-effective and versatile platform for studying marine ecosystems, monitoring ocean health, and gathering data on oceanic processes. Traditionally, marine research has been conducted using manned research vessels, submersibles, and buoys, which are expensive and often limited in their ability to cover vast areas of the ocean. Drones, however, offer greater mobility, flexibility, and affordability, enabling researchers to study marine environments more comprehensively.

In marine biology, drones are used to monitor marine species, including whales, dolphins, sharks, and sea turtles. Drones equipped with high-definition cameras or thermal sensors can track the movements of these animals, study their behavior, and monitor their habitats. This is particularly important for species that are endangered or difficult to track using traditional methods. Drones also allow scientists to gather data without disturbing the animals, minimizing the impact of human presence.

In addition to studying marine life, drones are used to monitor the health of coral reefs, which are vulnerable to climate change, ocean acidification, and pollution. Equipped with multispectral cameras, drones can assess coral bleaching, detect the spread of invasive species, and measure the impact of human activities such as fishing and tourism. This data is crucial for developing conservation strategies and protecting these vital ecosystems.

Drones are also employed in studying oceanographic processes such as currents, temperature, and salinity. By flying over the ocean or even flying into the water, drones can collect valuable data on ocean temperatures, water quality, and atmospheric conditions. This data contributes to the study of climate change, as the oceans play a critical role in regulating the Earth's climate.

The use of drones in marine research is also expanding in the field of ocean pollution monitoring. Drones can be equipped with sensors that detect pollutants such as oil spills, plastics, and chemicals. This data can be used to track the source of the pollution, assess its spread, and support cleanup efforts. Drones are particularly effective in monitoring large and inaccessible areas of the ocean, where pollution may go undetected by traditional means.

Despite their advantages, drones face several challenges in marine and oceanic research. Saltwater exposure, strong ocean currents, and adverse weather conditions can affect the performance and longevity of drones. Additionally, ethical concerns related to drone use in sensitive marine habitats must be addressed to ensure that drones do not disrupt ecosystems or harm marine life. However, as technology advances and drones continue to play a critical role in oceanic research, they are likely to become an indispensable tool in the study and preservation of our oceans.

Chapter 12: Drones in Urban Development

12.1 Infrastructure Inspection and Maintenance

Urban development increasingly relies on drones to conduct infrastructure inspections and maintenance. The application of drones in this field has revolutionized the way cities monitor and manage their physical assets, particularly large-scale infrastructures such as bridges, roads, railways, buildings, power lines, and utility networks. Traditional inspection methods often involve scaffolding, cherry pickers, or the labor-intensive process of manual inspection, all of which are costly, time-consuming, and pose safety risks to workers. Drones, by contrast, provide a safer, more efficient, and cost-effective alternative for performing these critical tasks.

The primary advantage of drones in infrastructure inspection is their ability to access hard-to-reach areas without the need for human intervention. Drones equipped with high-resolution cameras, infrared sensors, and LiDAR technology can fly over, under, or around structures, capturing detailed images and data that would be challenging to obtain using traditional methods. For example, bridges, which are typically inspected manually by workers climbing to difficult locations or using heavy machinery, can now be monitored by drones flying underneath the structure to inspect beams, cables, and supports. This not only improves efficiency but also reduces the risk of accidents.

In addition to improving safety, drones also help reduce downtime during inspections. Traditionally, infrastructure inspections require roads to be closed or other disruptions to be made to traffic or utilities. With drones, these disruptions are minimized, as drones can operate without halting transportation or requiring the erection of scaffolding. For instance, power line inspections, which typically require workers to climb poles or access remote areas, can be completed more quickly with drones, ensuring that maintenance or repairs are carried out without causing significant interruptions to the power grid.

Drones also allow for real-time data collection and immediate analysis, enabling maintenance teams to quickly identify issues and plan remedial actions. In some cases, drones can be used to detect early signs of wear or damage, such as cracks in concrete, corrosion on metal surfaces, or faulty electrical equipment, which may be invisible to the human eye. With high-definition cameras and thermal imaging, drones can pinpoint the exact location of defects, allowing for precise maintenance interventions that help prolong the lifespan of infrastructure and prevent costly, large-scale repairs.

The use of drones in infrastructure inspections is also helping to streamline the maintenance cycle. With regular drone inspections, cities can implement predictive maintenance strategies, identifying problems before they escalate into serious issues. This approach not only improves the longevity of assets but also saves cities money in the long term, as minor repairs are much less expensive than major overhauls.

Moreover, the data collected from drones can be archived and analyzed to identify long-term trends and patterns in infrastructure conditions. Over time, this data can provide valuable insights into the effectiveness of maintenance programs, the impact of weather on infrastructure, and the need for upgrades or replacements. In this way, drones are contributing to more data-driven, proactive approaches to urban infrastructure management.

Despite the benefits, the use of drones in infrastructure inspection does come with some challenges. One key challenge is the regulatory environment, as drone operations in urban areas often require permits and adherence to airspace restrictions. Furthermore, drone pilots need to be highly skilled to operate in complex environments, and equipment such as cameras, sensors, and GPS systems must be properly calibrated to ensure data accuracy.

12.2 Role in Smart Cities

The concept of smart cities is built upon the integration of technology into urban systems to improve efficiency, sustainability, and quality of life for residents. Drones play a significant role in the development and implementation of smart city technologies, serving as valuable tools for monitoring and optimizing various urban functions. In smart cities, drones contribute to the improvement of transportation systems, environmental monitoring, public safety, and energy management, among other aspects of urban life.

One of the most important roles drones play in smart cities is in monitoring and managing traffic. Drones equipped with high-definition cameras and sensors can provide real-time data on traffic flow, congestion, and accidents. This data can be used by city planners and transportation authorities to optimize traffic signals, redirect traffic, and improve road safety. In some cases, drones can also be used for crowd management in busy urban areas or during major events, helping city officials to monitor and manage public gatherings.

In the context of environmental monitoring, drones are essential for collecting data on air quality, temperature, and pollution levels. By flying over urban areas, drones can measure concentrations of pollutants such as nitrogen dioxide, sulfur dioxide, particulate matter, and carbon monoxide. This data helps city planners identify pollution hotspots, track changes over time, and develop policies aimed at reducing pollution and improving air quality. Drones can also be used to monitor the urban heat island effect, which occurs when cities experience higher temperatures than surrounding rural areas due to human activities and infrastructure. By mapping heat distribution, drones help city planners design more energy-efficient cities with better green spaces and sustainable architecture.

Moreover, drones contribute to the efficiency of public services in smart cities. For example, drones can be used to deliver essential supplies, such as medical or emergency equipment, to locations that are difficult to access by road. In areas with congested traffic or during natural disasters, drones can offer a faster and more reliable solution for delivering goods to residents or responders. Additionally, drones can assist in infrastructure inspections, as discussed earlier, allowing for efficient monitoring of vital city infrastructure, such as roads, bridges, and power lines.

Drones are also valuable tools for enhancing urban safety and security. In smart cities, drones are used to monitor public spaces, conduct surveillance in high-crime areas, and provide real-time information to law enforcement agencies. Drones equipped with cameras and thermal imaging can provide situational awareness during emergencies, such as fires, accidents, or terrorist attacks. Additionally, drones are being integrated into firefighting efforts, where they can be used to assess fire locations, monitor the spread of smoke, and help firefighters make decisions in real time.

The integration of drones into the smart city ecosystem is also closely tied to the development of the Internet of Things (IoT). IoT devices, such as sensors, cameras, and environmental monitoring tools, are connected to a centralized system that allows city officials to track and manage various aspects of urban life. Drones, as mobile devices equipped with their own sensors and cameras, can interact with this system to provide real-time data, enabling more effective management of resources and services. For example, drones can monitor the condition of public infrastructure, send alerts about weather changes or emergencies, and help track the location of traffic or waste collection vehicles.

As smart cities continue to evolve, drones will become increasingly important in enhancing urban operations and improving quality of life. They will play an integral role in creating sustainable, resilient, and efficient urban environments, where technology works seamlessly to benefit residents and the environment.

12.3 Urban Planning and 3D Mapping

Urban planning is an essential process in shaping the growth and development of cities. Drones have proven to be indispensable tools in urban planning by providing a detailed, accurate, and up-to-date view of the urban landscape. One of the primary uses of drones in urban planning is in the creation of 3D maps, which offer an in-depth understanding of the city's physical environment. These maps are useful for assessing land use, designing infrastructure, and simulating future development scenarios.

Drones equipped with high-resolution cameras and LiDAR sensors can capture vast amounts of data from the air, which can be processed to generate three-dimensional representations of buildings, streets, and other urban features. The precision of drone-derived 3D maps allows urban planners to assess the existing cityscape with great accuracy and identify areas in need of improvement. These maps also help in visualizing urban growth, as planners can simulate the impact of new development projects, such as the construction of roads, bridges, parks, or commercial centers, on the surrounding environment.

One of the primary advantages of using drones in urban planning is their ability to capture data in real time. Traditional surveying methods, such as ground-based surveys or satellite imagery, can take weeks or even months to gather, process, and analyze. Drones, on the other hand, can capture high-quality imagery and data within hours or days, enabling planners to make timely decisions based on the most current information available.

In addition to providing accurate topographical data, drones are also used to assess land use and zoning. Drones can capture aerial images of different urban areas, allowing city officials to monitor changes in land use over time. This is particularly important in rapidly developing cities, where land use patterns may change frequently. For example, drones can track the conversion of agricultural land into residential or commercial zones, providing valuable insights into urban sprawl and helping planners make informed decisions about where to allocate resources for development.

Drones are also used to monitor the progress of construction projects. By capturing aerial images of construction sites, drones can provide a comprehensive view of how projects are advancing, allowing project managers and city officials to track timelines, verify that construction is proceeding according to plan, and ensure that building codes and regulations are being followed. Drones can also identify potential issues on construction sites, such as safety hazards or delays, and help mitigate risks before they become more serious problems.

In combination with other technologies, such as geographic information systems (GIS) and 3D modeling software, drones have transformed the way cities plan for future growth. The ability to visualize urban areas in three dimensions enables planners to assess the potential impacts of new developments on existing infrastructure, public spaces, and the environment. Additionally, drone-collected data can be integrated with IoT systems, further enhancing the planning process by providing real-time insights into factors like traffic flow, air quality, and population density.

However, while drones offer numerous advantages for urban planning, there are challenges associated with their use. One of the main issues is the integration of drone technology with existing urban infrastructure. Cities need to invest in the necessary infrastructure to support drone operations, such as drone landing zones, charging stations, and regulatory frameworks. Additionally, concerns about privacy, security, and airspace management must be addressed to ensure that drone operations are conducted safely and responsibly.

12.4 Integration with IoT Systems

The Internet of Things (IoT) refers to the network of interconnected devices that communicate with each other and share data to improve efficiency and functionality. Drones, as mobile, data-collecting platforms, are increasingly being integrated with IoT systems to enhance their capabilities in urban environments. By connecting drones to a broader network of IoT devices, cities can enable seamless data flow, optimize urban systems, and improve the delivery of services to residents.

In an IoT-enabled urban environment, drones can collect data from various sensors embedded in infrastructure, such as traffic signals, waste bins, streetlights, and environmental monitoring stations. This data can then be transmitted to a central system for analysis, where it can be used to make real-time decisions. For example, drones could be used to monitor traffic conditions and send data to IoT-enabled traffic lights, allowing the system to adjust light timings to optimize traffic flow. Similarly, drones can assist in waste management by identifying full bins and notifying collection teams through an IoT system, ensuring that waste is collected efficiently.

One of the key benefits of integrating drones with IoT systems is the ability to provide real-time data to city officials, urban planners, and public service teams. For instance, drones used for infrastructure inspection can send real-time images and sensor readings to a centralized IoT platform, where analysts can assess the condition of structures and recommend maintenance actions. This allows for faster decision-making and more responsive urban management.

Drones are also being used to monitor environmental conditions, such as air quality, temperature, and humidity. These sensors can be connected to an IoT network, enabling cities to track pollution levels and identify trends in real time. This data can be used to inform policy decisions aimed at reducing pollution, optimizing energy consumption, and improving public health.

By connecting drones to IoT systems, cities can also enhance public safety and emergency response. Drones equipped with thermal imaging and cameras can monitor public spaces, detect heat signatures from fires or accidents, and send alerts to emergency responders through the IoT network. In the event of a disaster, drones can be used to assess damage, map affected areas, and guide rescue operations in real time.

12.5 Challenges in Urban Drone Deployments

While the potential for drones to contribute to urban development is vast, there are numerous challenges associated with their deployment in urban environments. One of the main challenges is regulatory compliance. Urban areas are often crowded, with complex airspace management needs. Drones must operate within established airspace regulations to ensure that they do not interfere with manned aircraft, helicopters, or other flying objects. Managing drone flights in densely populated areas requires careful planning and coordination with aviation authorities, city planners, and other stakeholders.

Another challenge is privacy and security. Drones equipped with cameras and sensors can capture vast amounts of data, including images and videos of private property and public spaces. Concerns about the potential misuse of this data for surveillance or monitoring individuals have raised ethical questions about the use of drones in urban settings. Ensuring that drones are used responsibly, with adequate safeguards to protect privacy and prevent abuse, is essential for gaining public trust.

Safety is also a significant concern when deploying drones in urban areas. Drones must be able to operate safely in complex environments with numerous obstacles, such as tall buildings, power lines, and pedestrians. Malfunctions, crashes, or collisions with other aircraft or objects pose risks to people and property. Cities need to implement robust safety protocols, including no-fly zones, flight restrictions, and emergency response plans, to mitigate these risks.

Infrastructure limitations present another hurdle to drone deployment. Urban environments require dedicated infrastructure to support drone operations, such as drone charging stations, landing zones, and data processing facilities. Additionally, urban drone operations must be coordinated with other technologies, such as traffic management systems and IoT networks, which can involve significant investment and integration efforts.

Despite these challenges, the benefits of drones in urban development are undeniable. With advancements in technology, regulatory frameworks, and public acceptance, drones will continue to play a pivotal role in shaping the future of cities, making them more efficient, sustainable, and responsive to the needs of their residents. As cities evolve and become smarter, drones will undoubtedly be key players in the ongoing transformation of urban life.

Chapter 13: Military and Defense Applications

13.1 Tactical Reconnaissance and Surveillance

The integration of drones into military and defense operations has transformed the way modern warfare is conducted. Tactical reconnaissance and surveillance have been two of the most significant applications of Unmanned Aerial Vehicles (UAVs) in military operations, playing an essential role in intelligence gathering, battlefield awareness, and force protection. UAVs, commonly known as drones, have provided a wide array of capabilities that allow military forces to collect valuable real-time information from both the air and the ground. The key advantage of UAVs in this domain lies in their ability to operate in environments that may be too dangerous or inaccessible for manned aircraft or ground personnel.

Tactical reconnaissance refers to the gathering of intelligence to provide commanders with situational awareness of enemy positions, movements, and activities. UAVs are equipped with various sensors, including high-resolution cameras, infrared sensors, and radar, that enable them to conduct reconnaissance missions over vast areas. These drones can operate in all weather conditions, day or night, and provide continuous surveillance over long durations without the need for refueling. By being able to fly at high altitudes, drones can observe large portions of the battlefield, giving military personnel a comprehensive view of the terrain, enemy troop deployments, and supply lines.

One of the key uses of UAVs for tactical reconnaissance is their ability to capture aerial imagery and video footage that can be transmitted back to military command centers in real time. This allows commanders to make informed decisions regarding troop movements, strategic positioning, and engagement. Drones can be deployed for missions to gather intelligence over areas that are too dangerous for human reconnaissance, such as hostile territories, combat zones, or enemy-held positions. The real-time data collected by drones enables quick decision-making and response, which is critical in the fast-paced nature of modern warfare.

Another significant benefit of UAVs in tactical reconnaissance is their ability to perform electronic warfare (EW) operations. Drones can be equipped with sensors to intercept communications or jam enemy signals, providing valuable intelligence on enemy operations. They can also be used to monitor radio frequencies, detect electronic signatures, and analyze electronic emissions. This capability gives military forces an advantage in disrupting enemy communications and impeding their ability to coordinate operations.

Additionally, UAVs are invaluable for surveillance operations, especially for monitoring enemy movements and identifying threats. Surveillance drones are often used in preemptive strikes to identify the presence of insurgents or enemy combatants in specific areas. By maintaining constant surveillance of key locations, drones can detect unusual patterns of activity or track suspicious movements, allowing military forces to respond swiftly and with precision.

UAVs are particularly effective for monitoring critical infrastructure such as bridges, roads, and communication facilities that may be targeted by adversaries. This monitoring capability helps ensure that vital infrastructure remains operational and that enemy attacks are detected early. Military UAVs can also serve as an early warning system by identifying potential threats such as approaching enemy forces or vehicles, providing time for defensive measures to be taken.

The integration of drones into military operations has also brought new technological advances to reconnaissance and surveillance. The use of artificial intelligence (AI) and machine learning (ML) algorithms has enabled UAVs to analyze large amounts of data in real-time, automatically identifying targets or anomalies that warrant further investigation. This has increased the efficiency of military operations, as drones can autonomously process surveillance data, reducing the reliance on human operators for routine tasks.

Despite the advantages, there are challenges associated with the use of UAVs in tactical reconnaissance and surveillance. Drones must be able to operate in contested environments, where they may be targeted by enemy air defense systems, electronic warfare tactics, or other countermeasures. Additionally, drones require reliable communication links to transmit data back to command centers, and this communication can be disrupted by enemy jamming or interception of signals.

13.2 Combat and Weaponized UAVs

Combat UAVs, often referred to as "drone strikes," have become one of the most controversial and strategically significant elements of modern military operations. Weaponized UAVs are specifically designed for carrying and delivering munitions, such as bombs, missiles, and guided projectiles, and they have been used in a wide range of combat scenarios, from targeted strikes against enemy forces to larger-scale military operations. These drones are used by several countries to conduct precise attacks in areas where the risks to human soldiers would be high or where traditional military assets, such as manned aircraft or ground forces, would be less effective.

Weaponized UAVs are typically equipped with a variety of payloads, including air-to-ground missiles, precision-guided bombs, and, in some cases, remotely operated machine guns or smaller firearms. The most widely known combat UAVs include the MQ-1 Predator and MQ-9 Reaper, both of which have been used by the United States in numerous military engagements. These drones provide an alternative to manned combat aircraft and offer several advantages over traditional methods of warfare.

One of the key benefits of weaponized UAVs is their ability to deliver precise strikes with minimal collateral damage. Unlike traditional bombing methods, which can affect large areas and cause civilian casualties, UAVs can be equipped with precision-guided munitions (PGMs) that allow for highly targeted strikes. These PGMs are capable of hitting specific targets, such as enemy combatants or military installations, with great accuracy. This capability is particularly important in counterterrorism operations, where the goal is often to eliminate specific individuals or groups without causing harm to nearby civilians or infrastructure.

Weaponized UAVs are also able to conduct long-duration missions, providing military forces with persistent surveillance and strike capabilities. The endurance of these drones, combined with their ability to remain in the air for hours or even days at a time, allows them to conduct continuous operations over vast areas. This is particularly valuable in regions such as the Middle East, where ongoing insurgent activities and conflicts require sustained attention.

Another advantage of combat UAVs is their ability to operate in high-risk environments with minimal risk to human life. By using drones to conduct strikes, military forces can avoid sending manned aircraft into dangerous airspace where they may be targeted by enemy fighters, surface-to-air missiles, or other threats. Combat UAVs allow for the neutralization of high-value targets without putting pilots or soldiers at risk. This has made UAVs a popular choice for conducting strikes in hostile or politically sensitive areas.

Additionally, UAVs offer the advantage of being relatively inexpensive compared to traditional manned aircraft, which require costly maintenance, pilots, and training. Weaponized UAVs are much less expensive to operate, making them an attractive option for countries with limited defense budgets. Their relatively low cost allows for more frequent missions, which is crucial in ongoing military operations that require continuous surveillance and strikes over extended periods.

Despite their many advantages, weaponized UAVs have raised significant ethical and legal concerns, particularly regarding the use of drones in targeted killings and military interventions. The use of drones in countries such as Pakistan, Yemen, and Somalia has sparked debate over the legality and morality of conducting strikes in foreign territories without the consent of local governments. Critics argue that the use of drones for extrajudicial killings violates international law and sovereignty, while others contend that drones provide an effective means of eliminating terrorist threats without endangering military personnel.

There are also concerns regarding the potential for civilian casualties in drone strikes. Although drones are highly accurate, there have been instances in which civilian deaths have resulted from targeting errors or misidentification of targets. These incidents have led to calls for greater transparency and accountability in drone operations, as well as improved oversight and control measures to ensure that drones are used in compliance with international humanitarian law.

13.3 Unmanned Logistics Support

Unmanned logistics support, another important application of drones in military operations, involves the use of UAVs to deliver supplies, equipment, and even ammunition to troops deployed in remote or hostile environments. Military forces are often tasked with maintaining supply lines in areas that are difficult to reach due to challenging terrain, weather conditions, or the presence of enemy forces. Drones are particularly well-suited to this role, as they can navigate through these obstacles and deliver critical supplies to troops on the front lines.

One of the most significant advantages of using UAVs for logistics support is their ability to operate in areas where traditional ground-based supply methods are impractical or too dangerous. For example, in mountainous or densely forested regions, it may be impossible or too risky to deliver supplies using trucks or other vehicles. UAVs can fly directly to designated locations, bypassing obstacles and enemy lines, to deliver food, medical supplies, ammunition, or equipment to soldiers in need. This capability is particularly useful in areas with limited infrastructure, such as conflict zones or disaster-stricken regions.

Drones can also be used to provide real-time monitoring of supply chains and inventory levels. UAVs equipped with sensors and cameras can track the movement of supplies from central depots to forward operating bases, ensuring that inventory is properly managed and that critical materials are always available when needed. This real-time visibility into logistics operations can help military forces avoid supply shortages and improve the overall efficiency of their operations.

Another advantage of UAVs in logistics is their ability to conduct resupply missions without the need for human personnel. This eliminates the risk to soldiers and drivers who would otherwise be exposed to enemy fire or ambushes during traditional resupply missions. UAVs can be flown remotely, or in some cases autonomously, to deliver supplies, allowing military personnel to focus on other tasks. Additionally, UAVs can be programmed to deliver supplies to multiple locations in a single mission, further improving efficiency and reducing the need for multiple resupply runs.

While unmanned logistics support provides significant benefits, there are challenges associated with its implementation. One key challenge is the limited payload capacity of most UAVs, which may restrict the amount of supplies that can be delivered in a single mission. While larger drones are capable of carrying heavier loads, they require more significant infrastructure and maintenance, making them less practical in certain environments. Another challenge is the need for secure communication and navigation systems to ensure that drones can operate safely and effectively in contested or hostile areas.

13.4 Counter-Insurgency and Anti-Terrorism Operations

Counter-insurgency (COIN) and anti-terrorism operations are two critical areas where UAVs have proven to be invaluable assets. In these operations, UAVs are used to gather intelligence, monitor enemy movements, provide surveillance over key targets, and deliver precision strikes against insurgent or terrorist groups. The ability of drones to operate in remote, hostile environments has made them an essential tool for modern military forces engaged in COIN and counter-terrorism efforts.

In counter-insurgency operations, UAVs are primarily used for intelligence, surveillance, and reconnaissance (ISR). Drones provide real-time data on the location of insurgent forces, their movements, and activities. This information can be used to identify patterns, assess threats, and plan strategic responses. UAVs equipped with high-resolution cameras, infrared sensors, and radar can detect enemy forces even in challenging conditions such as dense forests, mountainous terrain, or urban environments.

One of the key roles of UAVs in COIN operations is the identification of high-value targets (HVTs) – individuals or groups that pose a significant threat to national security. UAVs can track these individuals and gather intelligence on their activities, enabling military forces to plan precise and targeted strikes. By using drones to conduct surveillance over suspected insurgent hideouts or training camps, military forces can gather valuable information that leads to successful raids or strikes.

Drones are also used to disrupt insurgent supply lines, prevent the movement of weapons, and target enemy logistical operations. By monitoring enemy routes and facilities, UAVs can help prevent the flow of arms, equipment, and supplies to insurgent groups, weakening their operational capabilities. Additionally, drones can be used to gather intelligence on enemy recruitment efforts, providing valuable information on the structure and organization of insurgent networks.

In anti-terrorism operations, UAVs play a crucial role in identifying and eliminating terrorist threats. These operations often take place in regions where terrorist organizations are deeply embedded within local populations, making it difficult for traditional military forces to distinguish between combatants and civilians. UAVs offer a solution by providing a level of precision that is unmatched by other forms of military force. Precision-guided munitions and surveillance capabilities enable military forces to target specific terrorist cells or leaders while minimizing civilian casualties.

The use of drones in anti-terrorism operations has significantly increased over the past two decades, particularly in regions such as the Middle East and Africa. Drones have been used to target key terrorist leaders, disrupt training camps, and prevent the spread of terrorism. The ability to strike targets with pinpoint accuracy has made drones an essential tool in the fight against terrorism.

Despite the success of UAVs in COIN and counter-terrorism operations, there are challenges associated with their use. One of the primary concerns is the risk of civilian casualties, particularly in densely populated areas where insurgents or terrorists may be hiding. Strikes that result in unintended deaths or injuries can damage the legitimacy of military operations and fuel anti-government sentiment. As such, military forces must exercise caution and precision when using drones in COIN and counter-terrorism operations.

13.5 Ethical Concerns in Autonomous Warfare

The increasing use of autonomous UAVs in military operations has raised a host of ethical concerns, particularly in relation to the use of drones in warfare. These concerns revolve around issues of accountability, the potential for indiscriminate killing, the loss of human oversight, and the ethical implications of using machines to make life-and-death decisions.

One of the key ethical challenges associated with the use of drones in warfare is the question of accountability. When a drone conducts a targeted strike, it is often controlled remotely by an operator who is situated far from the battlefield. This raises questions about who is responsible if the strike results in unintended harm, such as civilian casualties or the targeting of the wrong individuals. The issue becomes more complicated when autonomous drones are involved, as the machines themselves may make decisions about when and where to strike, without human intervention. This lack of accountability can lead to concerns about the potential for mistakes or abuses of power in conflict zones.

Another ethical concern is the potential for drones to be used in an indiscriminate manner. Although drones are capable of conducting precise strikes, there is always the risk that a strike could harm innocent civilians or damage infrastructure. This issue is especially pronounced in areas with complex civilian populations, such as urban centers or conflict zones where insurgents may be operating among civilians. The use of drones in such environments requires careful targeting and surveillance to ensure that strikes are accurately aimed at combatants and do not harm innocent bystanders.

The loss of human oversight in autonomous warfare is another area of concern. As drones become more autonomous, they may be able to make decisions without human input, potentially leading to situations in which drones engage in combat without the moral and ethical considerations that guide human decision-making. The ability of machines to make decisions in life-and-death situations raises difficult questions about the ethical implications of replacing human judgment with algorithms.

Finally, there are concerns about the broader impact of autonomous drones on the nature of warfare. The increased use of drones in combat raises questions about the future of military engagements, including the possibility of conflict being waged without direct human involvement. Some critics argue that this could lead to a form of warfare in which decisions are made by algorithms, with no direct accountability to the people who are affected by the violence. This raises questions about the morality of fighting wars in which humans are removed from the immediate consequences of their actions.

In conclusion, while the use of drones in military and defense applications has brought numerous tactical and strategic advantages, it has also raised significant ethical challenges. The growing reliance on autonomous UAVs in warfare calls for a careful consideration of the moral, legal, and political implications of their use. Military forces must ensure that drones are used responsibly, with appropriate safeguards in place to protect civilians and uphold international humanitarian law. Additionally, the ethical concerns surrounding autonomous warfare require ongoing dialogue and debate to establish clear norms and principles for the responsible use of drone technology in military operations.

Chapter 14: Innovations in Drone Technology

14.1 Miniaturization and Swarm Robotics

The evolution of drone technology has seen dramatic advancements over the years, particularly in miniaturization and the development of swarm robotics. Both of these innovations are reshaping the way UAVs (Unmanned Aerial Vehicles) are designed, deployed, and utilized across various sectors, from defense to commercial applications. These advancements open up new possibilities for UAV capabilities, providing more versatile, efficient, and scalable systems.

Miniaturization of Drones

Miniaturization refers to the process of reducing the size of drone components while maintaining or even improving their performance. In recent years, there has been a substantial push to create smaller drones with enhanced power, efficiency, and functionality. This miniaturization is largely driven by advancements in microelectronics, battery technology, and lightweight materials.

One of the key benefits of miniaturized drones is their ability to operate in confined or difficult-to-reach spaces. These drones can be used in urban environments, search-and-rescue missions, or indoor applications where larger UAVs may struggle to navigate. Miniaturized drones can also be deployed in swarms, offering significant advantages in applications such as surveillance, reconnaissance, and environmental monitoring.

The technological progress in miniaturization has allowed drones to be equipped with sophisticated sensors, cameras, and communication systems despite their small size. These miniaturized systems are capable of performing complex tasks, such as capturing high-resolution imagery, detecting gas leaks, or conducting environmental monitoring. Moreover, the compact size of these drones enhances their mobility and maneuverability, allowing them to fly in tight spaces and avoid obstacles more effectively.

Miniaturized drones also benefit from reduced weight, which contributes to longer flight times and better energy efficiency. With less weight to carry, these drones consume less power, which can extend their range and endurance. Additionally, the miniaturization of components such as sensors, controllers, and motors leads to a reduction in the overall cost of drone production. This makes small drones more accessible to a wider range of industries, including agriculture, logistics, and security.

Swarm Robotics and Multi-UAV Coordination

Swarm robotics refers to the coordination of multiple drones working together to perform tasks collectively. Inspired by the behavior of insects such as ants or bees, swarm robotics enables drones to operate in concert, often without the need for centralized control. Each drone in the swarm operates autonomously but communicates with the other drones in the group, exchanging information to coordinate tasks efficiently.

The development of swarm robotics has had profound implications for a wide variety of applications. In the military and defense sectors, for example, drone swarms can be deployed for surveillance, reconnaissance, and targeted strikes. Swarm technology allows for the distribution of tasks among multiple drones, enhancing the system's redundancy and resilience. If one drone in the swarm fails, the others can take over its responsibilities, ensuring the mission continues without disruption.

In commercial applications, swarm robotics can be used in areas such as agriculture, environmental monitoring, and disaster response. In agriculture, for example, drone swarms can be used to monitor large fields, collect data on crop health, and distribute fertilizers or pesticides. The ability of drones to work in synchrony can dramatically increase the speed and efficiency of such tasks. Additionally, swarm systems are often designed to function with minimal human intervention, allowing for autonomous operation over long periods.

Swarm robotics also allows for highly flexible and scalable systems. As the need for drones increases, more UAVs can be added to the swarm, and their coordination can be managed through advanced algorithms. These algorithms allow the swarm to adapt to changing conditions in real time, making them highly dynamic and capable of performing tasks in challenging environments.

Despite the tremendous promise of swarm robotics, challenges remain in terms of communication, coordination, and security. The more drones in a swarm, the more complex the coordination becomes, requiring sophisticated algorithms and communication protocols to ensure that all drones operate in sync. Additionally, the risk of interference, whether from electronic jamming or cyber-attacks, presents potential vulnerabilities that must be addressed to ensure the integrity of swarm operations.

14.2 Advances in Battery and Power Systems

A critical factor in the success and evolution of drones is the continued improvement in battery and power systems. As drones become more advanced and are deployed for longer and more complex missions, the need for high-performance, long-lasting, and lightweight power sources becomes increasingly important. Battery technology, energy management systems, and alternative power sources are key areas where innovations are making significant strides.

Battery Technology and Improvements in Energy Density

The heart of most drones' power systems lies in their batteries. The energy density of batteries directly affects the drone's flight time, range, and overall performance. As drones become more sophisticated, their need for higher energy capacity has driven research into improving battery technology. Lithium-ion (Li-ion) and lithium-polymer (LiPo) batteries are the most common power sources for drones, but these technologies have limitations in terms of energy density and flight time.

Over the past decade, advances in battery chemistry have led to batteries with greater energy density, allowing drones to fly longer without adding significant weight. New materials, such as lithium-sulfur and solid-state batteries, are being developed to replace conventional lithium-ion batteries. These new battery types offer much higher energy densities, meaning that drones could potentially stay airborne for hours rather than just minutes. Additionally, solid-state batteries are safer, with less risk of fire or explosion, which is a significant advantage for drones used in military, industrial, or emergency scenarios.

Another emerging development in battery technology is fast-charging batteries, which significantly reduce downtime for drones between flights. Fast-charging capabilities are particularly important in commercial applications, such as delivery services or logistics, where drones need to be in operation for long hours with minimal breaks for recharging.

Alternative Power Systems and Hybrid Power Sources

While traditional batteries remain the most common power source for drones, alternative power systems are being explored to further extend flight times and improve performance. Solar-powered drones, for instance, use photovoltaic cells to harness solar energy during flight. These drones can recharge their batteries while in the air, significantly extending their operational endurance, especially in remote areas where recharging stations are not available. Solar-powered UAVs are ideal for long-duration surveillance, environmental monitoring, and scientific research missions.

In addition to solar power, hybrid power systems that combine fuel cells and batteries are also under development. Fuel cell-powered drones can operate for extended periods, as fuel cells provide continuous energy without the weight and limitations of conventional batteries. These systems work by converting hydrogen into electricity, producing a minimal environmental footprint. Hydrogen fuel cells have the potential to revolutionize drone technology, particularly for industrial and military applications that require long-range, high-endurance UAVs.

While alternative power systems offer numerous benefits, they come with their own set of challenges. Solar-powered drones, for example, depend heavily on weather conditions and may not perform well in cloudy or nighttime environments. Hybrid power systems, on the other hand, require complex integration and are often more expensive than traditional battery systems. Nonetheless, as technology continues to advance, it is expected that alternative power sources will become a more viable and widely adopted solution.

14.3 5G Connectivity and Drone Integration

The integration of drones with 5G networks has the potential to revolutionize drone technology by enabling faster, more reliable communication between drones, ground stations, and other connected devices. 5G connectivity offers significant improvements over existing cellular networks, including higher data transfer speeds, lower latency, and greater network capacity, making it ideal for enhancing drone operations across a wide range of applications.

Faster and More Reliable Data Transfer

5G networks provide a substantial boost in data transfer speeds compared to 4G or previous generations of cellular technology. This increase in speed allows for real-time transmission of large volumes of data, such as high-definition video feeds, sensor data, and telemetry information. For drones used in tasks such as surveillance, infrastructure inspection, or environmental monitoring, 5G enables seamless streaming of high-resolution video and detailed sensor data without the delays or interruptions that are common with older network technologies.

The low latency of 5G networks further enhances the responsiveness of drone operations. Latency, or the delay between sending and receiving data, is a critical factor for applications that require immediate decision-making, such as emergency response or autonomous drone operations. With 5G, the delay between the drone and the ground station is minimized, enabling operators to control drones more effectively and make real-time adjustments to flight plans or actions.

Enabling Autonomous and Swarm Drones

The combination of 5G connectivity and advanced AI algorithms opens the door for fully autonomous drones to operate safely and efficiently. 5G networks allow drones to communicate with each other and with central control systems, exchanging real-time data to coordinate actions. This is particularly important for swarm robotics, where multiple drones must collaborate on tasks such as monitoring a large area or conducting coordinated operations. With 5G, drones in a swarm can share data on their positions, sensor readings, and environmental conditions, improving coordination and optimizing task execution.

The enhanced communication capabilities offered by 5G also support the development of more advanced autonomous flight systems. Drones can receive continuous updates about their surroundings, helping them avoid obstacles, adjust flight paths, and respond to changes in real-time. In the context of autonomous navigation, 5G provides the low-latency communication necessary for making split-second decisions, ensuring that drones can operate independently and efficiently in dynamic environments.

Challenges of 5G Integration in Drone Operations

While the promise of 5G connectivity for drones is immense, there are challenges that need to be addressed. One of the primary concerns is the coverage and availability of 5G networks, particularly in rural or remote areas where drone operations are often critical. Additionally, the infrastructure required to support 5G, including base stations and signal towers, may be insufficient in some regions, hindering the widespread adoption of 5G-enabled drones. Security and privacy concerns are also important considerations, as increased connectivity increases the risk of cyber-attacks and unauthorized access to drone systems.

14.4 Biologically Inspired Drones

Biologically inspired drones, often referred to as bioinspired or biomimetic drones, are designed to mimic the physical characteristics and behaviors of living organisms. These drones take inspiration from nature, incorporating elements such as wing shapes, locomotion patterns, and sensory systems found in birds, insects, and other animals. The goal of biologically inspired drones is to create UAVs that are more efficient, maneuverable, and adaptable to diverse environments, offering capabilities beyond traditional drones.

Mimicking Flight and Locomotion

One of the most significant areas of biologically inspired drone design is flight mechanics. For example, researchers have studied the flight patterns of birds and insects to improve drone stability and maneuverability. Birds, such as falcons or hawks, are highly efficient fliers, using specialized wing structures to reduce drag and increase lift. By studying these creatures, engineers have been able to design drones with wings that mimic the flexible and adaptive movements found in nature.

Similarly, insects like bees and dragonflies exhibit remarkable agility, allowing them to hover, dart, and navigate complex environments. Researchers have developed micro-drones with insect-like flight capabilities, allowing them to hover in place, move with precision, and even fly in tight spaces. These drones are particularly useful in applications such as search and rescue, where navigating confined spaces or hovering in place for extended periods is necessary.

Biologically Inspired Sensing and Decision-Making

In addition to flight mechanics, biologically inspired drones also incorporate sensory systems that mimic those of animals. For instance, some drones are equipped with vision systems that replicate the wide field of view and depth perception found in certain species of insects or birds. These advanced vision systems enable drones to perceive their environment more accurately, detect obstacles, and identify targets or objects of interest.

The integration of biological principles into drone decision-making is also a growing area of research. Insects, such as ants, are capable of making collective decisions based on simple individual behaviors. Inspired by this behavior, bioinspired drones are being developed with decentralized control systems that allow them to make decisions autonomously without relying on a central command. This is particularly useful for swarm robotics, where drones must work together and make collective decisions in real-time.

14.5 Emerging Technologies in UAV Design

As drone technology continues to advance, several emerging technologies are shaping the future of UAV design. These innovations are pushing the boundaries of what drones can do, enhancing their capabilities in areas such as flight performance, autonomy, and environmental adaptability.

Quantum Computing and AI for Drones

Quantum computing, still in its infancy, holds immense potential for revolutionizing drone design. Quantum computers, which harness the power of quantum mechanics, are able to process large amounts of data far more efficiently than classical computers. This ability could be used to enhance the decision-making process for autonomous drones, allowing them to process sensor data, perform complex calculations, and adapt to dynamic environments in real time. The integration of quantum computing with AI algorithms could make drones far more intelligent, efficient, and capable of performing tasks with minimal human intervention.

3D Printing and Custom Drone Parts

The advent of 3D printing technology has the potential to change the way drones are manufactured and customized. 3D printing allows for the creation of complex, lightweight, and durable components that are tailored to specific design requirements. This enables more flexible, efficient, and cost-effective production of drones, particularly for specialized applications where traditional manufacturing methods may not be suitable. Drones can be rapidly prototyped, allowing engineers to test and iterate on designs in a fraction of the time it would take with conventional methods. Additionally, 3D printing allows for the creation of custom drone parts that can be easily replaced or upgraded, enhancing the versatility and longevity of UAVs.

Advanced Materials and Smart Fabrics

Emerging materials, such as smart fabrics and metamaterials, are also playing a key role in the evolution of UAVs. Smart fabrics can be integrated into drone designs to improve their aerodynamics, provide built-in sensors, or make the drones more adaptable to environmental changes. For instance, metamaterials that have been designed to manipulate light and sound could be used to reduce the visibility of drones, making them harder to detect by radar or other sensors.

Advanced materials also allow for lighter, more durable drone structures, improving overall performance and increasing payload capacity. As the field of materials science progresses, UAVs will continue to become more efficient, durable, and capable of operating in diverse environments.

In conclusion, innovations in drone technology are rapidly changing the landscape of UAV design, with developments in miniaturization, power systems, connectivity, biologically inspired designs, and emerging technologies contributing to a new era of capabilities. As these technologies continue to evolve, drones will become increasingly versatile, powerful, and integral to a wide array of applications, from commercial industries to defense and beyond.

Chapter 15: Regulations and Legal Framework

15.1 Global Drone Regulations Overview

As drones have rapidly evolved and found use across numerous sectors, the need for a robust legal framework has grown. The regulatory landscape surrounding drones varies significantly between countries and regions, and understanding this complex environment is crucial for safe, efficient, and lawful drone operations. Global drone regulations encompass various aspects such as airspace management, operator requirements, safety protocols, privacy concerns, and international cooperation.

The International Civil Aviation Organization (ICAO), a specialized agency of the United Nations, plays a key role in establishing globally recognized standards for civil aviation, which includes unmanned aerial systems (UAS). However, it is important to note that ICAO sets guidelines, while the responsibility for actual implementation rests with individual countries. This can result in differing regulations depending on the country's national policies, industry needs, and technological advancements.

In general, most countries have followed a similar approach in developing drone regulations, focusing on airspace control, safety, and privacy. The primary objectives of these regulations are to ensure the safety of manned aircraft, prevent accidents, protect privacy, and minimize the risks of drones being used maliciously. As of now, drone regulations are still evolving as technology advances, but a set of common regulatory elements can be identified.

One of the first steps in drone regulation was the creation of rules to ensure that drones are kept at a safe altitude and avoid manned aircraft. Restrictions on drone flight within controlled airspace are typical, and most countries have classified areas where drone operations are limited or prohibited, such as near airports or over populated areas. For example, in the United States, the

Federal Aviation Administration (FAA) has designated airspace classes (A to G), each with its own set of rules governing drone use.

Similarly, the European Union has established regulations under the European Union Aviation Safety Agency (EASA), which oversees the regulation of drones across member states. The EASA guidelines provide a framework for managing drone operations, including safety, training, and certification requirements. Other regions such as Asia and South America have also developed their own regulatory approaches based on local conditions and needs.

Despite efforts to create global consistency, significant variation exists. In some regions, drones are regulated as part of the existing aviation framework, while in others, new regulations are being created to address the unique challenges presented by unmanned flight. Countries such as the U.S., the EU, and China have been leaders in establishing comprehensive frameworks for drone regulation, while other countries may still be in the process of developing their policies.

With drones becoming a global phenomenon, international agreements and frameworks are also being pursued to create standardization. The aim is to allow cross-border drone operations, making it easier for operators to use drones internationally without navigating conflicting regulations. Although ICAO's guidance provides a foundation, it is likely that significant coordination among nations will be needed to ensure that drone operations are consistent across borders.

15.2 Airspace Management and Integration

One of the most critical aspects of drone regulations revolves around airspace management and integration. As drones become increasingly popular, they share the skies with manned aircraft, leading to complex challenges regarding airspace allocation, safety, and communication. Effective airspace management is necessary to ensure that drones do not interfere with commercial, military, or emergency flights.

Challenges in Airspace Integration

The integration of drones into existing air traffic control systems has proven to be one of the most complicated regulatory challenges. Traditionally, air traffic management systems have been designed around manned aircraft, and adapting these systems to account for UAVs has required significant changes. Drones typically fly at lower altitudes than most commercial aircraft, which creates the potential for collisions if they are not adequately separated.

A major issue is that many drones operate autonomously or semi-autonomously, meaning there is no pilot onboard to monitor the aircraft's surroundings. This lack of situational awareness increases the risk of accidents, especially in crowded or busy airspaces. Therefore, integrating drones into air traffic management (ATM) systems requires the development of technologies such as detect-and-avoid (DAA) systems, which allow drones to detect other aircraft and take evasive action if necessary.

Another challenge is that drones are often small and hard to detect by traditional radar systems, especially if they are flying at lower altitudes or in rural areas. To mitigate these risks, some countries require that drones be equipped with identification systems, such as Remote Identification (Remote ID) in the U.S., which transmits the drone's location and other relevant information to authorities. This allows air traffic controllers and law enforcement to track drones and respond to potential safety issues in real time.

Airspace management strategies must also account for the differing operational environments of various drone types. For example, while some drones may operate within confined spaces or at very low altitudes (such as in urban environments), others may be used for long-distance flights, requiring higher altitudes and integration into broader airspace sectors. Additionally, drone swarms present a unique challenge, as managing multiple autonomous drones requires coordination mechanisms that ensure safe and efficient flight paths without congestion or interference.

U-Space and UTM Systems

In Europe, the concept of U-Space is being developed to address airspace management specifically for drones. U-Space refers to a set of services and procedures designed to ensure safe and efficient drone operations in low-level airspace (below 150 meters). It includes systems for tracking, communication, and coordination between drones and manned aircraft. The goal of U-Space is to provide a safe and scalable airspace environment for drone operations, including the use of advanced communication technologies and automation.

In the U.S., the FAA has been developing its own Unmanned Aircraft System Traffic Management (UTM) system. UTM is a framework for managing drone operations, focusing on ensuring safe and efficient coordination between drones and other airspace users. The FAA is also working with private industry partners to develop technologies that will enable drones to communicate with one another and air traffic control systems, further enhancing safety and integration.

As the number of drones in the skies increases, airspace management systems will need to evolve to accommodate the growing demand for drone operations. The goal is not just to ensure safety but also to promote efficient use of airspace. With the implementation of systems like U-Space and UTM, it is expected that airspace congestion will be minimized, and drones will be able to operate safely alongside manned aircraft in shared airspace.

15.3 Data Privacy and Security Concerns

The rapid proliferation of drones has raised significant concerns regarding data privacy and security. Drones equipped with cameras, sensors, and other monitoring technologies have the ability to capture large amounts of data, including video footage, geospatial data, and personal information. As drones are used for a variety of applications, from surveying and surveillance to scientific research and commercial operations, the risk of data breaches and privacy violations has become a pressing issue.

Privacy Risks and Public Concerns

One of the primary privacy concerns surrounding drones is the potential for unauthorized surveillance. Drones equipped with high-resolution cameras and other sensing devices can easily capture images and videos of private property, public spaces, and individuals without their consent. This has raised alarms among privacy advocates, who argue that drones could be used to infringe on personal privacy rights or gather sensitive information without proper oversight.

To address these concerns, several countries have implemented regulations that limit the use of drones for surveillance. For example, in the European Union, the General Data Protection Regulation (GDPR) provides legal protections regarding the collection, use, and storage of personal data, including data collected by drones. Under the GDPR, drone operators must comply with strict guidelines regarding data handling, including ensuring that personal data is not captured unnecessarily and that any data collected is securely stored and used for legitimate purposes.

In addition to privacy concerns, drones also pose security risks. Drones can be hacked or intercepted, potentially allowing malicious actors to take control of the UAV or misuse the data being collected. As drones become more connected to the internet and other communication networks, the potential for cyberattacks increases. To mitigate these risks, security protocols such as encryption, secure communication channels, and authentication systems are being integrated into drone systems.

Regulations Addressing Privacy and Security

Many governments have enacted laws that address privacy and security concerns associated with drone operations. In the U.S., for example, drone operators are prohibited from using UAVs for surveillance in a manner that violates an individual's reasonable expectation of privacy. Similarly, in the U.K., the use of drones for surveillance is regulated by the Surveillance Camera Code of Practice, which sets out rules for the deployment of drones by public authorities.

In addition to these regulations, there are ongoing efforts to establish more comprehensive legal frameworks for data protection in the context of drone operations. For instance, some countries have developed specific laws governing the use of drones for commercial and governmental surveillance, which include requirements for transparency and accountability in data collection. These laws often require drone operators to inform individuals if their data is being collected and provide an option to opt-out or seek redress if privacy rights are violated.

15.4 Licensing and Operator Training Requirements

As drones become more integrated into various industries, it is increasingly important to ensure that operators are properly trained and licensed to operate them safely and legally. The complexity and risks involved in drone operations—particularly in commercial and industrial contexts—have led to the establishment of training and certification programs for drone operators.

Licensing Programs and Operator Certification

In many countries, drone operators must obtain specific licenses or certifications before they are allowed to fly commercially. For example, in the United States, the FAA requires commercial drone operators to obtain a Remote Pilot Certificate, which involves passing an aeronautical knowledge test and demonstrating proficiency in drone operation. Similarly, in the

European Union, operators must hold a Remote Pilot Certificate issued by an approved training organization before flying drones for commercial purposes.

The licensing process typically includes both theoretical and practical training. Theoretical training covers topics such as airspace classification, flight safety, weather conditions, and regulations governing drone use. Practical training, on the other hand, focuses on the hands-on operation of drones, including flight techniques, navigation, and emergency response procedures.

Specialized Training for Advanced Operations

For more advanced drone operations—such as those involving large drones, complex flight paths, or high-risk missions—additional training and certification may be required. For example, in the U.S., commercial drone operators wishing to perform operations such as flights over people, night operations, or operations in controlled airspace must obtain a waiver from the FAA, and in some cases, undergo additional training. Similarly, operators of drones used for industrial inspections, infrastructure monitoring, or emergency response may require specialized training to ensure they are proficient in handling the specific equipment and challenges associated with those tasks.

Continuing Education and Re-certification

As drone technology evolves rapidly, ongoing education and re-certification are becoming increasingly important. Drone operators need to stay current with the latest regulations, safety protocols, and technological advancements in the field. Some countries have established requirements for drone operators to complete continuing education courses and renew their licenses periodically. This ensures that operators are aware of new regulations and best practices in the rapidly changing drone landscape.

15.5 Balancing Innovation with Public Safety

The rapid advancement of drone technology presents significant opportunities for innovation across a wide range of sectors, from delivery services to infrastructure inspections and environmental research. However, the increasing ubiquity of drones also raises concerns about public safety, airspace congestion, privacy violations, and the potential misuse of UAVs. The challenge for regulators and policymakers is to strike a balance between fostering innovation and ensuring that drones are operated safely and responsibly.

Fostering Innovation in Drone Technology

Innovation in the drone industry is essential for driving progress in various sectors, and regulators recognize the need to create an environment that encourages technological advancements. For instance, regulators have begun to explore ways to allow more flexibility in drone operations, such as expanding the permissible scope of operations for commercial drone pilots, allowing operations in controlled airspace, and permitting beyond-visual-line-of-sight (BVLOS) flights under certain conditions.

Innovative technologies such as artificial intelligence, machine learning, and autonomous flight systems have the potential to revolutionize the drone industry. However, ensuring that these technologies are deployed safely requires careful consideration of their implications. For instance, autonomous drones must be able to safely navigate through complex environments and handle unexpected obstacles. Additionally, the introduction of drone swarms—where multiple drones work in tandem to perform a task—raises questions about coordination, safety, and airspace management.

Ensuring Public Safety and Security

While fostering innovation is important, public safety must remain a top priority. Regulations should be developed to address potential risks posed by drones to both individuals and society at large. This includes preventing collisions with manned aircraft, protecting privacy, securing sensitive data, and ensuring that drones are not used maliciously for surveillance or attack purposes.

As drones are deployed in an increasing number of industries, it is also crucial that their operators are well-trained and adhere to stringent safety standards. The development of robust safety protocols, such as no-fly zones, geofencing, and fail-safe systems, can help mitigate risks and reduce accidents.

In conclusion, finding the right balance between fostering innovation and ensuring public safety is a challenging but essential aspect of drone regulation. With the right regulatory frameworks in place, the potential benefits of drones can be fully realized while minimizing risks and addressing public concerns. As the drone industry continues to grow, regulators will need to remain agile, continuously updating laws and policies to accommodate new technologies and emerging applications.

Chapter 16: Challenges in Drone Technology

16.1 Technical Challenges in Drone Development

The development of drone technology, though rapidly advancing, is far from being free of challenges. A host of technical issues continues to affect the progress of the industry, influencing not only the design and functionality of drones but also their safety, reliability, and overall effectiveness. Many of these challenges arise from the complex interplay between the multiple components of a drone, each of which requires meticulous engineering to ensure smooth operations. Understanding these technical hurdles is key to recognizing the potential and limitations of drone technology.

Power and Battery Life

One of the primary technical challenges in drone development is the issue of power, particularly in relation to battery life. Drones rely heavily on lightweight batteries to power their motors, sensors, and other essential systems, yet the energy density of current battery technologies limits the amount of time a drone can remain airborne. While advances in lithium-polymer (LiPo) batteries have improved the efficiency of drones, these batteries still provide relatively short flight times, typically ranging from 20 to 60 minutes depending on the size and weight of the drone.

This limited flight time is especially problematic for commercial applications, such as long-range delivery services, infrastructure inspections, or agricultural monitoring, where drones need to cover large distances or work for extended periods without recharging. The power-to-weight ratio of batteries is a critical factor, and despite ongoing research into more efficient energy storage technologies, the industry is still constrained by battery limitations. Innovations in fuel cells, solar-powered drones, and alternative energy sources are actively being explored, but widespread adoption remains a distant goal.

Flight Stability and Control Systems

Another significant challenge in drone development is ensuring stable flight and effective control systems. While many drones are equipped with advanced flight controllers, sensors, and stabilization systems, maintaining stability in various conditions remains difficult. Factors such as wind, turbulence, and the weight distribution of payloads can negatively affect a drone's performance. Particularly in challenging environments such as urban settings, dense forests, or open seas, drones must be able to adjust in real-time to changing conditions, which requires advanced algorithms and robust control systems.

Drones that rely on autonomous systems for flight need to make rapid decisions based on real-time data collected by sensors such as accelerometers, gyroscopes, GPS, and barometers. These sensors, while essential for stability, can be prone to errors, particularly in environments with GPS signal interference, such as urban canyons, or in situations where visual sensors may not function properly due to fog, dust, or darkness. The development of more reliable and redundant control systems is critical to improving the safety and robustness of drones.

Autonomous Navigation and Collision Avoidance

As drones continue to operate in increasingly complex environments, the challenge of autonomous navigation and collision avoidance becomes more pressing. While drones are capable of operating autonomously for most standard operations, their ability to navigate unknown environments and avoid obstacles remains a significant hurdle. Technologies such as LiDAR, ultrasonic sensors, and computer vision have been integrated into drones to enable real-time obstacle detection, yet no system is infallible. The rapid development of machine learning algorithms that enable drones to "learn" from their surroundings and adapt to dynamic situations is essential for overcoming this challenge.

However, developing a system that not only detects but also responds to obstacles in a timely and efficient manner, especially when operating in congested environments such as cities, is far more difficult than it might seem. The increased computational power required for these systems, along with the need for real-time data processing and decision-making, presents an additional layer of technical complexity.

Miniaturization and Integration of Components

As drones become smaller and more capable, the challenge of miniaturizing components and integrating them into a compact form becomes increasingly important. Many of the most advanced drones on the market today are built with a wide range of sophisticated components, including sensors, cameras, flight controllers, and communication devices. The challenge lies in integrating these components into a lightweight, efficient package that does not compromise on performance.

The miniaturization of electronics, especially the sensor arrays required for high-level autonomy and data collection, is an ongoing challenge. Miniaturization often results in reduced power output or compromised accuracy, requiring constant innovation to strike a balance between size, weight, and functionality. Furthermore, achieving better integration of software and hardware in drones, without creating systems that are too complex or prone to failure, is a persistent challenge.

16.2 Weather and Environmental Limitations

Weather and environmental conditions are a fundamental constraint in drone operations. While drones are capable of flying in many different settings, various external factors such as wind, rain, snow, temperature extremes, and fog can significantly affect their performance. These environmental limitations, which are often outside the control of drone operators, must be considered in both the design and operation of drones.

Wind and Turbulence

Wind is one of the most common environmental factors that can affect drone performance. Strong winds can cause a drone to lose stability, making it difficult for the operator to control the flight path or maintain a safe altitude. Drones, particularly small consumer models, are especially vulnerable to high winds and gusts, which can make them difficult to fly, prone to drifting, or even lead to crashes. While larger drones equipped with more powerful motors can generally withstand higher wind speeds, many still have upper limits of wind resistance, making them unsuitable for use in extreme conditions.

Turbulence, especially in urban areas, can also pose a significant challenge. As drones fly at lower altitudes, they can encounter turbulent air caused by buildings, terrain, and other obstacles, which disrupt smooth airflow and increase the likelihood of instability. The development of better sensors and flight algorithms to account for and adapt to turbulent conditions is critical in addressing these challenges.

Rain, Snow, and Ice

Weather events such as rain, snow, and ice present additional difficulties for drone operations. Drones are typically vulnerable to water damage, particularly in their electronics and sensors. Rain, for instance, can short-circuit components, impair the performance of cameras or sensors, and increase drag, making it harder for drones to fly efficiently. Additionally, snow and ice buildup on drone propellers or body can cause significant reductions in flight performance and control, leading to a loss of stability and, in some cases, mechanical failure.

While many commercial drones are designed with some level of weatherproofing, it is challenging to ensure that drones remain operational under all conditions. The development of specialized drones that are capable of operating in extreme weather conditions, such as those used in disaster relief or search and rescue operations, is an active area of research. These drones may feature rugged designs, weather-resistant coatings, or even specialized heating elements to prevent the buildup of snow or ice.

Temperature Extremes

Temperature extremes can also affect drone performance. Drones are generally optimized for operation within a narrow temperature range, and exposure to high or low temperatures can have significant effects on battery life, motor efficiency, and sensor accuracy. In cold environments, battery efficiency tends to decrease, reducing flight time and overall power. On the other hand, in hot environments, drone components, including the motor and electronics, may overheat, leading to reduced performance or failure.

Visibility and Fog

Low visibility conditions, such as those caused by fog, haze, or smoke, are particularly problematic for drones relying on visual sensors or GPS. Drones with cameras and other optical sensors can struggle to accurately perceive their surroundings in foggy conditions, leading to issues with navigation and obstacle detection. Similarly, GPS signals, which many drones rely on for positioning, can become weak or unreliable in areas with significant obstructions, such as dense forests or urban canyons.

To address these environmental limitations, companies are developing drones with advanced sensor suites, including radar, LiDAR, and infrared sensors, which can work in low-visibility environments. These technologies, though promising, come with their own set of challenges in terms of cost, integration, and efficiency.

16.3 Issues with Scaling for Commercial Use

As drones move from niche applications to widespread commercial use, a series of scaling issues must be addressed. These include challenges related to mass production, operational efficiency, regulatory compliance, and cost-effectiveness.

Mass Production and Standardization

Scaling drone manufacturing for commercial use often requires standardization, which can be difficult given the wide variety of drone types and configurations needed for different applications. Manufacturers must develop cost-effective and scalable production processes while maintaining the quality and performance of each unit. Additionally, standardizing drone components, such as batteries, sensors, and motors, across various models can lead to cost reductions but may limit the customization that some commercial sectors require.

Infrastructure for Drone Operations

For large-scale commercial use, drones require supporting infrastructure, including systems for charging, maintenance, and monitoring. The widespread adoption of drones for applications like parcel delivery or surveillance necessitates the development of infrastructure that can support fleets of drones, as well as integrate drones into existing logistics and transportation networks.

For example, package delivery drones would need a network of designated takeoff and landing zones, as well as automated systems for managing deliveries, coordinating between drones, and ensuring timely and efficient operations. The logistics of managing large fleets of drones, including air traffic control systems, maintenance depots, and battery charging stations, presents a significant scaling challenge.

Cost-Effectiveness

The cost of drones, particularly those designed for commercial use, remains a barrier to widespread adoption. Although the price of consumer drones has decreased significantly in recent years, commercial-grade drones capable of heavy payloads, long-range flights, or complex operations remain expensive. Scaling up production, while reducing costs without compromising on safety and performance, is a key challenge for the industry.

16.4 Public Acceptance and Trust

Despite the immense potential of drones, gaining public acceptance remains a significant challenge. Concerns about safety, privacy, noise, and the general impact of drones on daily life have led to skepticism and resistance in some communities. Building public trust in drone technology is critical for its widespread acceptance and integration into society.

Safety Concerns

One of the most significant factors influencing public opinion about drones is safety. High-profile incidents involving drones, such as crashes, near-misses with manned aircraft, and incidents where drones interfere with emergency operations or public events, have contributed to public fears. Ensuring that drones are safe to operate and that operators adhere to strict safety protocols is crucial for addressing these concerns.

Privacy Issues

Drones equipped with cameras, sensors, and other data-gathering tools raise significant privacy concerns. The potential for drones to capture personal data or conduct surveillance without consent has led to calls for stricter privacy regulations. People are concerned that drones could be used to spy on private properties, monitor public spaces without oversight, or collect data for commercial or governmental purposes without their knowledge.

Governments and regulatory bodies are exploring ways to address these privacy concerns by creating guidelines for data protection, establishing no-fly zones over private properties, and requiring drone operators to obtain explicit consent before conducting surveillance activities. However, balancing the need for privacy with the benefits of drone technology remains a complex issue.

Noise Pollution

Drones, especially those with loud motors or those operating in large numbers, can contribute to noise pollution in urban and rural areas. As drones become more ubiquitous, concerns about their impact on quality of life—particularly in residential areas—are increasing. The development of quieter drone models, as well as the regulation of drone noise levels, is an ongoing effort to address these concerns.

16.5 Addressing Ethical and Privacy Concerns

Ethical considerations are central to the debate surrounding drone technology. Drones raise a host of questions about the responsible use of technology, particularly when it comes to surveillance, security, and warfare. Ensuring that drones are used ethically and in compliance with legal frameworks is essential for maintaining public trust and ensuring that the technology benefits society as a whole.

Surveillance and Privacy

As drones become increasingly capable of capturing high-resolution images and video, they are likely to be used more frequently for surveillance purposes. While this has legitimate applications in sectors like law enforcement, disaster management, and search and rescue, it also raises concerns about individual privacy. Without clear guidelines and regulations, drones could be used to monitor individuals or entire populations without consent, leading to a potential violation of privacy rights.

Autonomous Warfare

The development of weaponized drones and the potential for autonomous drones to engage in military operations raises profound ethical concerns.

Autonomous drones, capable of identifying and targeting individuals without human intervention, challenge traditional notions of accountability, responsibility, and decision-making in warfare. Ethical concerns about the use of drones in military operations, particularly in terms of civilian casualties, surveillance, and the potential for abuse, must be addressed by policymakers and international organizations.

In conclusion, while drone technology holds immense promise, its widespread adoption faces significant technical, ethical, and social challenges. Overcoming these hurdles requires ongoing research, thoughtful regulation, and collaboration between manufacturers, governments, and the public. Addressing these challenges will be crucial to unlocking the full potential of drones and ensuring their responsible and beneficial use across a range of industries.

Chapter 17: Future of Drone Technology

17.1 Trends in Autonomous Systems

The evolution of autonomous systems is one of the most transformative trends shaping the future of drone technology. The demand for autonomous drones is driven by their ability to operate without human intervention, thus enhancing their utility across various sectors. These drones are becoming more sophisticated, relying on artificial intelligence (AI), machine learning, and sensor technology to make real-time decisions in increasingly complex environments. The future trajectory of autonomous drones promises to redefine industries such as delivery services, agriculture, surveillance, and disaster management.

One of the most significant trends in the development of autonomous systems for drones is the move towards fully autonomous navigation and operation. Currently, drones are primarily semi-autonomous, requiring human intervention for various tasks such as takeoff, landing, and emergency responses. However, with advances in AI and deep learning, drones are increasingly capable of performing tasks entirely on their own, with minimal or no human oversight. This development is particularly relevant in sectors where human safety is at risk, such as in firefighting, search and rescue, or military reconnaissance.

Autonomous drones will be able to make decisions based on a vast array of inputs. These include environmental data from onboard sensors (such as cameras, LiDAR, and radar), information about the drone's position and surroundings, and external data from cloud-based systems or ground control stations. The integration of real-time data processing through edge computing will allow drones to process this information swiftly and make accurate decisions on flight paths, obstacle avoidance, and mission completion.

Another key trend is the development of swarm technology. Swarming refers to the ability of multiple drones to work together autonomously to complete complex tasks that a single drone would find challenging. Swarms of drones can be used for large-scale data collection, infrastructure inspection, or even logistics operations. For example, a group of drones can be deployed to map an entire city or conduct large-scale environmental monitoring, with each drone performing specific tasks while maintaining communication and coordination with others. This is achieved through advanced algorithms that enable real-time synchronization of tasks among the drones.

Moreover, autonomous drones will increasingly use predictive analytics and artificial intelligence to optimize their performance. With the help of AI, these drones can learn from their surroundings, improving their flight efficiency, decision-making capabilities, and ability to handle unforeseen circumstances. In the future, drones may even be able to predict maintenance needs or identify performance anomalies before they become significant issues.

As autonomous drones become more sophisticated, regulatory and safety considerations will need to evolve. Fully autonomous drones will require robust systems for collision avoidance, fail-safes in the event of communication failure, and advanced methods for airspace integration to ensure they can safely coexist with manned aircraft. Additionally, ensuring the privacy and security of data collected by these drones will become a critical concern, particularly as drones become more pervasive in surveillance and monitoring roles.

17.2 Potential in Space Exploration

While drones are already revolutionizing industries on Earth, their potential in space exploration is equally exciting and full of promise. The application of drones in space exploration could help overcome many of the challenges that space missions currently face, especially in terms of remote exploration, data collection, and the autonomy of robotic systems. As space exploration becomes more ambitious with missions to Mars, the Moon, and beyond, drones are poised to play an increasingly significant role in gathering data, mapping terrain, and carrying out critical tasks in the absence of human astronauts.

Drones could be essential for planetary exploration, especially on planets and moons where the harsh environment makes it difficult for rovers or manned missions to operate efficiently. For example, drones could be used to scout areas that are difficult to reach by other means, such as high cliffs, deep craters, or vast open plains. In the case of Mars, NASA's Ingenuity drone, which flew on the surface of Mars alongside the Perseverance rover, is a proof of concept for the potential of aerial vehicles in space exploration. Ingenuity demonstrated that drones could fly in the thin atmosphere of Mars, capturing invaluable data and providing unique perspectives of the Martian surface.

In future space missions, drones could be used to survey and map areas on the surface of other planets and moons. These drones could be equipped with advanced sensors and cameras to capture high-resolution imagery and conduct environmental analyses. For instance, drones could be employed to analyze surface compositions, detect mineral deposits, and assess the atmosphere of distant planets. This data could be used to guide further exploration and identify areas that may be of interest for human colonization or resource extraction.

The future of space exploration could see autonomous drones working in tandem with other robotic systems, such as landers and rovers. In this scenario, drones would provide aerial surveillance of the terrain, helping rovers navigate difficult environments or identify locations for further analysis. They could also act as relay stations for communication, improving the flow of data between surface operations and mission control.

The integration of drones into space missions could lead to significant advancements in how humanity explores space. Drones could be used not only for planetary exploration but also for monitoring spacecraft, inspecting satellites, and conducting maintenance activities. Autonomous drones could be deployed in low-Earth orbit to assist in repairing or servicing satellites, which would reduce the need for costly and risky human spacewalks.

17.3 Role in Global Connectivity Initiatives

As global connectivity remains a key driver of economic growth, drones are set to play a central role in bridging the digital divide, especially in remote and underserved regions. With their ability to provide high-speed internet access in areas lacking traditional infrastructure, drones could significantly enhance global communication networks. This would be particularly important in rural or disaster-stricken areas, where the installation of traditional telecommunications infrastructure is often impractical or cost-prohibitive.

One of the most notable initiatives in this space is the development of internet-delivering drones, which are designed to fly in the stratosphere, providing internet connectivity to remote or underserved regions. These drones are equipped with high-capacity communication equipment that can beam internet signals over large areas, much like satellites, but at a fraction of the cost. By operating in the lower stratosphere (typically 20 to 30 kilometers above the Earth), these drones can provide broadband coverage to areas that are out of reach of traditional ground-based infrastructure.

For instance, companies like Alphabet's Loon project (formerly known as Project Loon) have explored the use of high-altitude balloons and drones to deliver internet access in remote parts of the world. These high-altitude platforms can stay in place for long periods, creating a reliable communication network that can be used for education, healthcare, business, and emergency services. In regions affected by natural disasters, where traditional communication infrastructure is often damaged, drones could offer a rapid and cost-effective solution for restoring connectivity and facilitating communication between affected populations and aid organizations.

Furthermore, drones could be used in combination with 5G technology to enhance connectivity in urban areas. The use of drones to deploy 5G infrastructure in crowded environments could ease network congestion, reduce latency, and increase bandwidth availability for users. This would be particularly important as the demand for data-intensive services, such as virtual reality, autonomous vehicles, and IoT devices, continues to grow. Drones can act as mobile base stations, providing on-demand 5G coverage to high-density areas like stadiums, airports, or urban centers.

The future of global connectivity through drones could also extend to the provision of IoT services. IoT devices, which rely on seamless connectivity for data transmission, could benefit from the mobile, flexible nature of drone networks. Drones could serve as relay points for IoT data, providing a stable communication link between sensors in remote or hard-to-reach locations and central data hubs.

17.4 Emerging Applications in Healthcare and Education

The future of drones also holds significant promise in the healthcare and education sectors, where their ability to provide timely access to critical resources, data, and services can revolutionize care delivery and learning opportunities. Drones have already made an impact in areas such as medical supply delivery, remote diagnostics, and emergency medical response, and their role is expected to expand further in the coming years.

Healthcare Applications

Drones are increasingly being used to deliver medical supplies, such as vaccines, blood, and medications, to remote or hard-to-reach locations. In countries with limited infrastructure, particularly in Sub-Saharan Africa or mountainous regions, drones have already been deployed to carry life-saving medical supplies over long distances, bypassing roadblocks or difficult terrain. This ability to deliver medicine and medical equipment quickly, even in disaster-stricken areas, can save lives and improve healthcare access for underserved populations.

In addition to supply delivery, drones may also play a role in the transportation of diagnostic equipment. For example, drones could be used to carry medical devices to remote areas for diagnostic purposes or to transport patient samples to laboratories for testing. The speed at which drones can deliver these resources is particularly important in situations where time is of the essence, such as in the case of blood transfusions or pathogen detection.

The use of drones for telemedicine is another promising area of healthcare innovation. Drones can facilitate remote consultations by delivering diagnostic equipment or telemedicine kits to patients in rural areas. These kits could include portable ultrasound devices, thermometers, stethoscopes, and other tools, allowing healthcare professionals to remotely diagnose patients and prescribe treatments.

Education Applications

Drones also hold great potential for transforming the educational landscape. In remote areas, drones could be used to deliver educational materials, textbooks, and even learning devices to students in underserved regions. In addition to delivery, drones could be employed as educational tools in classrooms, offering hands-on learning experiences in subjects like science, geography, and engineering. Students could learn how drones work, their applications in various industries, and how they are integrated into larger technological systems.

Drones are also useful in educational settings that require environmental data collection. For example, students in environmental science programs can use drones to study ecosystems, monitor pollution levels, or collect meteorological data, all in real-time. Drones can be equipped with sensors that provide valuable insights into the environment, and students can use these data to make observations, conduct experiments, and perform research.

17.5 Vision for the Next Decade

Looking ahead, the next decade promises to be an exciting period for the drone industry, characterized by rapid technological advancements, wider adoption, and integration into both everyday life and industrial operations. Drones will become more autonomous, more capable, and increasingly interconnected. The future will see drones play an even more critical role in improving efficiency, productivity, and quality of life across multiple sectors.

The key developments over the next decade will likely include:

1. **Integration of AI and Machine Learning**: Drones will be able to make more intelligent decisions, adapt to changing environments, and operate independently in complex scenarios.
2. **Widespread Commercial Adoption**: Drones will be used extensively in industries like agriculture, logistics, construction, and healthcare, becoming standard tools in many professions.
3. **Global Connectivity**: Drones will play a key role in bridging the digital divide, providing global internet access to remote and underserved populations.
4. **Ethical and Regulatory Evolution**: As drones become more ubiquitous, there will be an increased focus on creating ethical frameworks and regulations that ensure their safe and responsible use.

In the next decade, drones are poised to make a profound impact on global society, offering innovative solutions to age-old challenges while presenting new opportunities for technological and industrial growth.

Chapter 18: Case Studies and Success Stories

18.1 Drones in Disaster Management: Real-World Applications

Disaster management is one of the most vital areas in which drones have proven their worth, and their real-world applications have had a transformative impact on the way emergency response teams operate. Drones are increasingly being utilized for a variety of disaster scenarios, including natural disasters like hurricanes, earthquakes, floods, and wildfires, as well as human-made crises such as industrial accidents or search-and-rescue operations. The rapid response capability of drones, combined with their ability to access areas that are difficult for humans to reach, has made them indispensable tools for disaster management.

A notable example of drones in disaster management occurred during the 2017 hurricane season in Puerto Rico. After Hurricane Maria devastated the island, drone technology played a crucial role in delivering real-time aerial imagery of the affected areas. Emergency teams used drones to assess the extent of damage to infrastructure, including bridges, roads, and buildings. The drone footage provided emergency responders with high-resolution maps that helped them prioritize areas for rescue operations. In this case, drones helped save lives by identifying trapped people in remote areas and facilitating the delivery of aid.

Another example comes from the aftermath of the 2015 Nepal earthquake, which caused significant destruction across the region. In this case, drones were deployed to conduct aerial surveys of the earthquake-affected areas. They were able to quickly assess the damage to infrastructure, particularly in remote or difficult-to-access locations, and provide authorities with valuable information to guide their relief efforts. Drones were also used to deliver supplies to hard-to-reach locations and transport medical teams to remote areas that were cut off from traditional transportation routes.

In the case of wildfires, drones have proven invaluable for monitoring fire behavior, assessing risk zones, and guiding firefighting efforts. During the California wildfires in 2020, drones were employed to survey vast areas of forest, providing real-time heat maps to help firefighting crews determine which areas were at the highest risk of spreading flames. This aerial intelligence allowed firefighting teams to be more efficient in their containment efforts, thus preventing further damage and saving lives. Drones equipped with thermal imaging sensors played a critical role in detecting hot spots and fire progression, which could easily have been missed by ground-based teams.

Additionally, drones have been used for delivering emergency medical supplies, such as blood, vaccines, and other life-saving equipment, in disaster-stricken areas. This is particularly important when traditional supply chains are disrupted due to damaged roads or infrastructure. In Rwanda, for example, the use of drones to deliver blood supplies has become a regular part of the healthcare system, ensuring that hospitals in remote areas receive the supplies they need even in the event of natural disasters or pandemics.

In these real-world disaster management scenarios, drones have demonstrated their ability to improve the speed, efficiency, and accuracy of emergency responses. They have provided valuable insights, saving time, resources, and lives, while enabling responders to make informed decisions. Moving forward, the role of drones in disaster management will likely grow, with advancements in drone technology allowing for faster deployment, greater operational range, and increased automation.

18.2 Transformative Impact on Industries

Drones are rapidly reshaping a wide range of industries, from agriculture and construction to logistics and telecommunications. The ability to collect high-resolution data, monitor progress in real-time, and operate in difficult environments has unlocked new possibilities for these industries. Drones are no longer just tools for photography or surveillance; they have become essential components of industrial operations, driving efficiency, safety, and cost savings.

In agriculture, drones have had a particularly transformative impact. Precision farming, which involves using data to optimize the efficiency of agricultural practices, has been greatly enhanced by drone technology. By utilizing drones equipped with cameras and sensors, farmers can monitor crop health, assess soil conditions, and identify potential problems, such as pest infestations or nutrient deficiencies, much earlier than traditional methods allow. The real-time data provided by drones can help farmers make informed decisions on irrigation, fertilization, and pesticide use, thereby reducing costs and increasing yields.

A real-world example of the transformative impact of drones in agriculture comes from the use of drone-powered crop spraying. In Japan, for instance, drone spraying systems are now commonly used to apply pesticides and fertilizers. Drones are able to cover large areas quickly and with precision, ensuring that crops receive the correct amount of treatment without the waste typically associated with traditional methods. This has not only improved crop productivity but has also minimized the environmental impact of pesticides, as drones can apply the chemicals only where they are needed, reducing overuse.

In construction, drones are being used for site surveys, progress monitoring, and inspections, all of which contribute to greater project efficiency and safety. Construction companies use drones to create detailed 3D models of construction sites, helping to plan layouts, track progress, and identify potential issues early on. For instance, a major construction firm in Dubai used drones to monitor the progress of the construction of the Dubai International Airport expansion project. By flying drones over the site, the company was able to collect aerial imagery and generate 3D maps, which allowed for better planning, reduced delays, and more accurate cost estimations.

Drones are also playing a key role in logistics and supply chain management, particularly in the field of last-mile delivery. In the e-commerce industry, companies such as Amazon and UPS are testing drone-based delivery systems to reduce delivery times and costs. Drones have the potential to revolutionize the logistics industry by providing faster, more efficient delivery solutions. The ability to deliver packages directly to customers in urban or rural locations without relying on traditional road infrastructure could greatly reduce the time and cost involved in shipping.

Additionally, in telecommunications, drones are being used to improve network coverage and connectivity. Companies like AT&T have begun using drones to inspect cell towers, reducing the need for human workers to climb dangerous heights and inspect equipment. Drones equipped with high-definition cameras and sensors can identify issues with antennas, cables, and other components, enabling telecom companies to perform proactive maintenance and ensure uninterrupted service.

Drones are also being utilized in the energy sector, particularly for inspecting power lines, wind turbines, and solar panels. Companies in the oil and gas industry use drones to inspect pipelines and offshore platforms, preventing costly downtimes and improving safety by keeping workers out of dangerous environments. By performing inspections via drones, companies can reduce labor costs, minimize risks, and gather more accurate data, leading to better decision-making and optimized operations.

18.3 Startup Ecosystems and Innovations in UAVs

The rapid growth of the drone industry has led to the emergence of a vibrant startup ecosystem that continues to push the boundaries of innovation in unmanned aerial vehicles (UAVs). From improving drone hardware to developing new software applications and business models, startups are at the forefront of transforming drone technology. These innovations have fueled the rise of drones across various industries and are helping to address existing challenges while creating new opportunities for commercial and industrial applications.

Startups in the drone sector are developing innovative drones that are smaller, more efficient, and equipped with advanced sensors and AI capabilities. One example of this innovation is the development of hybrid drones, which combine the capabilities of both fixed-wing and rotary-wing drones, offering the flexibility of vertical takeoff and landing (VTOL) with the endurance and efficiency of fixed-wing flight. These drones are ideal for applications like surveillance, monitoring large areas, and surveying remote locations.

Other startups are focused on improving drone autonomy. By integrating machine learning, computer vision, and edge computing into their UAVs, these companies are developing drones that can fly autonomously, avoid obstacles, and adapt to changing environments. One such startup, Skydio, has made significant strides in autonomous drone technology, offering drones that are capable of flying through complex environments without the need for a human pilot. These drones are equipped with advanced sensors that allow them to detect and avoid obstacles in real-time, making them ideal for applications in industries such as construction, agriculture, and public safety.

The software ecosystem surrounding drones is also rapidly evolving. Startups are developing platforms that allow businesses to manage fleets of drones, process large amounts of data collected by UAVs, and even analyze that data using artificial intelligence. Companies like DroneDeploy and senseFly have created drone mapping and surveying software that enables users to generate 3D models, perform volumetric analysis, and analyze aerial imagery with minimal effort. This has been a game-changer for industries such as construction, agriculture, and mining, where accurate geospatial data is essential for decision-making.

Furthermore, many startups are working on innovative drone-based services. For instance, some companies are developing drone-based delivery systems, while others are focused on using drones for infrastructure inspection, environmental monitoring, and emergency response. These startups are not only driving technological innovation but also creating entirely new business models and revenue streams.

The startup ecosystem in the drone industry is also fostering collaboration with larger corporations, research institutions, and government agencies. These partnerships are helping to accelerate the development and deployment of drones in various sectors, including agriculture, energy, and logistics. With more funding and attention being directed toward drone startups, the industry is likely to see continued growth and innovation in the coming years.

18.4 Lessons from Failed Projects

While the drone industry has seen tremendous success in various applications, there have also been notable failures and setbacks that offer valuable lessons for the future. Failed projects can help identify areas where improvements are needed, whether it's in drone design, operational processes, or regulatory compliance. By understanding the reasons behind these failures, businesses and governments can develop better strategies and mitigate the risks associated with drone technology.

One of the most significant challenges in the early days of drone deployment was the issue of regulatory compliance. Many companies launched drone-based services without fully understanding the legal framework governing drone operations, leading to a series of high-profile failures. For example, in 2015, Amazon's ambitious Prime Air project, which aimed to deliver packages via drones, faced significant delays due to regulatory hurdles in the United States. Despite the potential of the technology, the FAA's strict regulations on drone flight paths, altitude restrictions, and operational limitations forced Amazon to reconsider its approach. This case highlighted the importance of collaborating with regulators early in the development process to ensure that drone services can be deployed safely and within the confines of the law.

Another area where failures have occurred is in the development of autonomous drones. While the idea of fully autonomous UAVs holds great promise, the technology is still in its infancy, and many early attempts at creating drones capable of fully autonomous flight ended in failure. Issues such as poor sensor accuracy, lack of real-time decision-making capabilities, and problems with obstacle avoidance contributed to the setbacks in this area. The inability of drones to safely operate in complex environments led to crashes and loss of confidence in the technology. However, these failures have spurred rapid advancements in AI, computer vision, and sensor technology, helping to improve the reliability and safety of autonomous drones.

Another example of a failed project comes from the commercial drone delivery space. Several companies, including Google and Uber, initially pursued ambitious drone delivery systems, but they faced challenges related to logistics, public safety, and cost. These projects often struggled to scale effectively, as the infrastructure needed to support large-scale drone deliveries was either inadequate or non-existent. Additionally, drones faced significant opposition from communities concerned about noise pollution, privacy, and safety. These setbacks have taught the industry valuable lessons about the importance of building public trust, creating effective infrastructure, and addressing regulatory challenges before rolling out new drone services.

18.5 Inspiration for Future Developments

Despite the challenges and occasional failures, the drone industry is still poised for significant growth and innovation. The lessons learned from past mistakes are shaping the development of the next generation of drones, helping companies and researchers refine their technologies and strategies.

One of the key lessons from past failures is the importance of building strong partnerships between drone companies, regulators, and communities. Ensuring that drones are deployed safely, ethically, and in compliance with local regulations is essential for gaining public trust and fostering widespread adoption. Additionally, the collaboration between industries and startups is crucial for driving innovation, allowing for the sharing of resources, expertise, and ideas.

The rapid advancements in AI, sensor technology, and battery systems provide a strong foundation for the future of drone technology. As these technologies continue to evolve, drones will become more capable, autonomous, and versatile. Future developments may see drones capable of performing complex tasks such as precision medical surgeries, real-time data analysis in environmental monitoring, and even autonomous transportation services.

Inspiration for future drone applications can also be drawn from the industries and sectors that have already embraced drone technology. From agriculture and logistics to healthcare and disaster management, the widespread use of drones in these fields offers valuable insights into how drones can solve real-world problems and create new opportunities. As drone technology continues to evolve, there is a growing potential for UAVs to become an integral part of everyday life, whether it's in assisting with healthcare delivery, improving urban mobility, or advancing scientific research.

The future of drones is undoubtedly exciting, with continued innovation and the development of new, transformative applications. By learning from past experiences, understanding the current challenges, and fostering collaboration between key stakeholders, the drone industry is poised to make a lasting and meaningful impact across the globe.